# No Time for Playing Cards

# NO TIME
## FOR PLAYING CARDS

HIGHER RELIABILITY ORGANIZING FOR THE FIRE SERVICE

DANE CARLEY

CRAIG NELSON

Professional Education Products PennWell®

Fire Engineering®
BOOKS & VIDEOS

**Disclaimer**

The recommendations, advice, descriptions, and the methods in this book are presented solely for educational purposes. The author and publisher assume no liability whatsoever for any loss or damage that results from the use of any of the material in this book. Use of the material in this book is solely at the risk of the user.


1421 South Sheridan Road
Tulsa, Oklahoma 74112-6600 USA
800.752.9764
+1.918.831.9421

sales@pennwell.com
www.FireEngineeringBooks.com
www.pennwellbooks.com
www.pennwell.com

Marketing Manager: Sarah De Vos
National Account Manager: Cindy J. Huse

Director: Matthew Dresher
Managing Editor: Marla Patterson
Production Manager: Sheila Brock
Production Editor: Tony Quinn
Compositor: TIPS Technical Publishing, Inc.
Cover Designer: Charles Thomas

Library of Congress Cataloging in Publication Data Available on Request



Printed in the United States of America

1 2 3 4 5     21 20 19 18 17

# CONTENTS

# PREFACE

We wrote this book because we have a passion for seeing people and departments in the fire and emergency services be successful. Our passion for the fire service motivates us to improve it. We want to ensure that the vital services we provide to our communities continue to be provided safely, successfully, and effectively for generations to come.

In many ways, the emergency services represent the best of our society. Emergency responders take great pride in serving their community. Responders do not ask about a person's beliefs, values, or political positions because we are sworn to help them in their times of need regardless of sex, race, color, creed, socioeconomic background, or anything else. Emergency services exist solely to help society.

The techniques, methods, and processes that follow are based on not only our own research and data but also those from many other professions and industries interested in being the best. Doing things the way we always have just because we have always done them that way is no longer sufficient. From now into the future, we will be required to have a good reason for doing things the way we do—whether because it is faster, cheaper, safer, or more effective. This is not merely being suggested by our boards, commissions, communities, and media; it is now being required. They—and our own members, too—deserve to know the decisions being made are based on more than a gut feeling; they must be backed up by research and data. However, this does not mean that we need to do all the research ourselves or reinvent the wheel. Often, we simply need to review our own past experiences and study the successful methods of others to figure out which methods will be successful for us.

Success requires continual improvement because there is never an end point (unless you will be closing your fire department's doors). It requires a constant assessment of your surroundings and how your operations sync with those surroundings. The Earth will continue to spin, and change will always be present (just like death and taxes). The part that is often misinterpreted is the urgency with which changes must sometimes be enacted. Either we do not see the problem coming or we see it but do not realize the negative impact it will have until it is too late. We play a victim of uncontrollable circumstances or assume that the way we do things today will still be correct tomorrow. For example, when it comes to budget cuts, there were often indicators months to years in advance of the actual cuts. The longer indicators are ignored, and the longer you wait to make necessary changes, the harder it will be to overcome their impact.

How we approach and handle change is often a good indicator of success. Change is not fun or easy for anyone; however, the most successful people and departments understand that it is better to effect change than to let it affect them. Change is going to happen, so you may as well have some say in how it happens. At best, taking a proactive approach to change may avert negative outcomes, but more likely you will at least be able to lessen the negative impact. For example, budget cuts might still occur, but their impact could be diminished.

Improvement is accomplished by identifying gaps in effective service. These gaps are often found by asking why and using data to find the root cause. To that end, there is always new data to follow, especially now that we can access just about anything we want to record. Ignoring the data will result in eventual failure—meaning poor service, increased injuries and deaths, budget cuts, and poor decision making. The reality is that you will never do less—unless your department is in trouble.

In our society and in our emergency services, we are drawn to make decisions based on gut feelings and emotions, both of which are often independent from reality and far from correct. Experienced gut reactions are good at an emergency scene because they allow us to react to dynamic situations through recognition-primed decision making; however, the majority of our time is not spent at an emergency scene, and recognition-primed decision making is not as useful in nonemergency situations. Nevertheless, many of our decisions are still based on gut reactions and feelings. These decisions are not necessarily tied to how to best serve the public but instead to what helps us the most. We must always remember that we exist to serve the citizens first, not ourselves, and our decisions should reflect that priority.

Is this book about data? Yes and no. Mostly it is about human behavior. Still, data provides a way to quantify human behavior and human desires (your community's definition of valuable service) to help find solutions. For example, the Underwriters Laboratories Fire Service Research Institute vent studies tell us that our gut feeling to ventilate a fire can be counterproductive because we may not be able to provide a big enough vent to release the heat faster than the fire is producing it.

This book is about organizational and individual leadership, it is about personally striving for improvement for every member of a department at all levels, and it is about learning to ask why. Some people may say, "You're trying to change an entrenched culture—a culture that is good for the service and that shouldn't be changed," or, "You are just trying to change for the sake of changing." On the contrary, we have decades of fire service experience between the two of us and feel strongly that the culture has many positive attributes that we would like to see institutionalized so they are not lost. For example, emergency services personnel are critical thinkers with great problem-solving skills and a never-give-up attitude; these traits should continue to be institutionalized in a positive way. Having said that, we would also like to see the service develop behaviors that cause us to evaluate certain attributes of our own culture that are or may become detrimental to either our own safety or the public's safety, detract from service to our community, or pose an unnecessary burden to our constituency with the same vigor. In short, we would like to see a department be the best it can be by constantly learning—by constantly asking why. We strive for a point in time when we no longer hear, "Because we've always done it that way" in answer to a question of why. Sometimes, that may mean tweaking our culture, but that is not always bad (and is how we achieved many steps in our evolution so far, including powered apparatus, SCBA, radios, and, more generally, increased professionalism). We would like departments to use those critical thinking skills internally (but positively) for self-evaluation, to find ways to improve on our indispensable problem-solving skills.

We are excited about the future of fire and emergency services. We think the people who are part of the fire and emergency services are the best people in the world—people who want to be the best at what they do. We hope that reading this not only provides you with the tools to improve but also excites and encourages you. We hope you will strive to improve yourself, your department, and your profession!

## How this book is formatted

This book relies heavily on examples and reader interaction. This is done in several ways.

- ***Case in Point***: This may or may not be a real-life example, but it is helpful to illustrate the point being made.
- ***Real-Life Example***: These are case studies taken directly from the National Fire Fighter Near-Miss Reporting System. Because each represents someone else's experience in their own words, we did not change their punctuation, grammar, or spelling, although brackets are occasionally used for clarification and to denote reviewer de-identification. Each is divided into an event description and the lessons that person learned from the experience. We do make certain assumptions about the incident when asking questions to tie the incident to the topic being discussed.
- ***Discussion Questions***: These questions are designed to tie the examples to the topic being discussed in some cases. In others, they are designed to prompt readers to reflect on themselves or their departments.
- ***Exercises***: These are also designed to tie the examples to the topic or to prompt readers to reflect on themselves or their departments by practicing a skill.

We have written this book for many uses. We hope that it is easy to sit down and read on your own time, but we hope that it can be used as a training tool as well. Whether it is used in a formal training setting or with your crew, we have included the examples and exercises so that it helps you, the reader, or so that you can take it to your crew or a class and facilitate discussion.

# ACKNOWLEDGMENTS

Writing this book has been a great opportunity for us that would not have been possible without the support and understanding of our wives, Paula Nelson and Alicia Carley, and our children, Dane and Kjersten Nelson and Gunnar, Cade, and Teague Carley. It does not seem like saying thank-you carries enough weight for all that you have done to support us while we were working on this project. But we will start with that: Thank you!

We are also grateful to our peers who were willing to spend time reviewing our work and providing great suggestions. We would like to thank:

- Ron Guggisberg—Captain, Fargo, ND, Fire Department
- Jason Underdahl—Captain, Fargo, ND, Fire Department
- Rick Loveland—Firefighter, Moorhead, MN, Fire Department and Fire Program Coordinator at Minnesota State Community and Technical College in Moorhead
- Jeff Wallin—Assistant Chief, Moorhead, MN, Fire Department

And last but certainly not least, we would like to thank the staff at PennWell Publishing for encouraging us, giving us a chance, and giving us a platform to reach the fire service by writing this book and being patient with us while we learned about the publishing process. They have not only given us a chance but have become friends over the years. Had Bobby Halton not called our radio show one night to encourage us, this would have likely never happened. Thank you for the push, Bobby. We have made many friends at PennWell Publishing and Fire Engineering over the years but must specifically thank Diane Rothschild, Pete Prochilo, and Mike McEvoy who have each played a key role in helping us over the years.

1

# WHY WE DO WHAT WE DO

## CHAPTER OBJECTIVES

- Detail how external forces can influence internal decisions
- Explain how those decisions can affect departments
- Describe ways to reduce the effect of external influences by understanding human factors

## THE PEOPLE MACHINE

Human behavior drives decision making, influencing every single decision being made by every person, in every department, on every day. Obvious as that may seem, it is rarely acknowledged in the training of emergency responders. As a group, we have our own behaviors and tendencies that affect the decisions we make. For example, 68% of emergency responders are boredom-susceptible sensation seekers (Jensen 2005). This means we tend to like action and a little bit of danger—but in a "sensible" way.

### Case in Point

Picture someone riding a motorcycle at over 100 mph down the highway with no helmet or leathers. The firefighter or EMT would think, "What an idiot! There goes another organ donor." Ironically, that same emergency responder, being a boredom-susceptible sensation seeker, is likely to do the same thing but think it is safe, having chosen to wear a helmet and leathers. Firefighters enjoy taking some level of risk as long as it is perceived by them to be safe.

Regardless whether we admit it, people tend to be predictable. Habits built by a department's culture, training, and practices cause firefighters as a group to react similarly. Besides teaching firefighting skills, the goal of fire academies and training is to reinforce a strong work ethic: it makes us dependable, skillful, and prideful. Importantly, this training enables us to do our jobs well, but it also causes everyone to think similarly. Ultimately, the fire service attracts people with similar personalities to begin with and then puts them through training that makes them even more similar. Studying human factors allows us to identify the positive and negative behaviors developed during this process. Understanding this enables a department to reinforce positive behaviors and provide training to correct negative behaviors.

## THE HUMAN FACTOR IN A DYNAMIC ENVIRONMENT

Awareness of human factors dates to the Industrial Revolution. This awareness developed because people were having accidents with machines that caused injuries, loss of life, and substantial property damage. Although these machines had become bigger, faster, and more complex as time passed, the people operating them did not experience the same degree of evolution. Thus, as humans began to interact more with machines, it became apparent that machine design must account for human interaction.

Airplanes, being nothing more than flying machines, are a good example of this evolution. Over time they grew larger, faster, and more complex. The airplane cockpit also exemplifies that machine and human interactions may have little room for errors or omissions, as well as the potential for serious consequences when the interactions fail. Eventually, because flying machines crashed at an unacceptable rate, investigators sought the root causes of these accidents and determined that most of the time there was little or nothing mechanically wrong with the airplanes. This left the human side of the equation, and a conclusion of these investigations was that humans operating in highly dynamic, complex, and hazardous environments are likely to make mistakes. Thus, human factors became a strong focus in pilot training—a focus that continues to this day, more than 30 years later.

Emergency scenes are also dynamic, complex, and hazardous environments in which responders, who are human, interact. The work environment for firefighters and EMTs is in many cases even more complex than that within the airplane cockpit. We respond to situations where something has already gone wrong, and we generally do not have a lot of detailed information about the problem. Yet, training on human factors, while emerging, remains almost nonexistent in typical emergency services training (both the fire service and EMS).

Over time, the science of human factors has grown to encompass other areas of manufacturing, processing, engineering, design, military, and government industries in an effort to boost productivity and increase safety. Machines are considerably safer and more reliable; nevertheless, accidents do occur in spite of these developments, in large part because humans still have to interact with machines, and the highest cause of accidents today is the human factor.

Consider car accidents, for example. How often is the primary cause of an accident involving a car (a complex machine) mechanical in nature? Now speculate about how often the person operating the car was distracted (talking on the phone, texting, eating, applying makeup, shaving). Many other nonmechanical factors—such as poor lighting, alcohol use, and poor weather conditions—contribute to accidents. The point is that the car did not get itself into the accident—a human did. Therefore, integrating human factors into a root-cause analysis of accidents is a key component in development of a learning culture—that is, a culture constantly looking for improvement.

In general, emergency responders have difficulty dealing with the human side of the equation. This may be because humans are complex, so true solutions must likewise be complex. Much of our knowledge of the human side of the equation is relatively new, having been learned only in recent years. Some departments are already taking strides in this area by incorporating the human part of the equation into their training and operations. They have realized that although it takes more work and study, it is also the area where we have the most room for improvement in the emergency services.

### Discussion Question 1–1

What traits do you have in common with your fellow firefighters? Do not focus on political beliefs or upbringing but instead on basic behaviors and interests, such as a habit or part of the culture (e.g., risk taking).

## WHAT LIMITS AFFECT HUMANS?

Human beings have limits. As skilled as all firefighters and EMTs are, we encounter situations that overload our own capabilities, mental and physical, leading to errors.

As part of our teaching and research, we have reviewed numerous Mayday incidents. Many of those fire incidents involved clearly audible Mayday calls that were missed simply because the incident commander was overloaded.

Research has shown that once your pulse rate reaches 150 beats per minute, your auditory processing ability is lost (you can't hear). Even though this is primarily a physical limitation, it is clear that not hearing could affect your decisions on the fire ground. Moreover, one person cannot do everything without becoming mentally overwhelmed. Human factors helps us understand ourselves so that we can recognize situations that will likely exceed our own limits.

## WHAT LIMITS THE BRAIN'S ABILITIES?

The brain is an amazing organ capable of doing things in ways the medical community is just beginning to understand. But it does have some limitations relevant to the fire service, which include:

- The inability to multitask
- Constantly filtering all the information and stimuli in the surrounding environment
- The tendency to find preferential information that supports a preconceived notion or bias
- The tendency to react to incoming information before it is completely processed
- The tendency to go with the flow
- Stress

### Doing two things at once—or not

Firefighters interact with a variety of dynamic environments. For example, we think of driving a fire engine to an emergency incident as a relatively simple task because we do it often. However, behaviorists studying human factors would consider this a complex situation with multiple tasks occurring simultaneously. Firefighters drive large vehicles at high speeds, operating lights, sirens, and air horns; maneuvering through traffic; planning routes; recalling building access; and spotting the vehicle—all while watching to make sure the other drivers react appropriately and do not add a further hazard.

This is a complex, dynamic process. Our senses work together to make it seem like we are multitasking to accomplish all of these things at once, but in reality our brains switch between each of our senses up to 60 times per second to give us that illusion (Putnam, n.d.). By understanding this aspect of human factors, you can appreciate that when many events occur simultaneously, you are more likely to miss important cues because your brain's attention is being divided. For instance, the driver might be focusing on a car that is making questionable actions and consequently miss the captain's instruction about a wrong turn.

When a person reduces the number of distractions to the brain, the other senses get more attention. For example, the loss of one's eyesight may seem to result in better hearing. In reality, though, the auditory sense simply gets more attention because it is not competing with eyesight (Putnam, n.d.).

## Discussion Questions 1–2

Has an incident ever overwhelmed you because of a need to complete multiple tasks at once? How did you handle it?

## Exercise 1–1

To practice multitasking, pick a destination in your response area. Then recite the most efficient route to that destination while juggling. If you cannot juggle, instead occupy your mind and hands by intubating a mannequin or changing an SCBA bottle while reciting the route.

### Not how my mother did it

The human mind likes equilibrium (for everything to make sense); a lack of equilibrium creates stress. In other words, the human mind wants what it is seeing and hearing to fit neatly into a bigger picture, and the mind may go to great lengths to fit the information it has into an existing belief system. This tension between beliefs and information is called cognitive dissonance, and the human mind shapes information to correct this disequilibrium. For this, it turns to biases (the way your mind would like it to be) to fit the pieces into the puzzle. The mind's biases are so strong that it will ignore obvious but conflicting data.

What the human mind sees in the present is influenced by what that person has believed for years because the brain seeks to balance current input with long-held values and beliefs. These values and beliefs are built throughout a person's life, which is why these influences are so strong. We see this in the fire service. The ongoing debates between the use of solid bore and fog nozzles or between vertical ventilation and positive-pressure ventilation (PPV) provide just two examples. None of those techniques are right 100% of the time. However, many of the arguments supporting these positions are based on a relatively small number of personal experiences and very little unbiased data. A firefighter who supports the use of a fog stream from a combination nozzle will almost always remember the personal, anecdotal evidence supporting that position but may forget the personal, anecdotal evidence refuting the position. Therefore, that firefighter will almost always grab a combination nozzle in a fog pattern because in his or her mind that is the most appropriate tool for the job.

### Discussion Question 1–3

When you debate, have you caught yourself arguing about something you have little background in but feel a strong emotional attachment to?

## Counting to ten

Has anyone ever told you to count to ten before responding to what someone else said? The premise behind this is that even though it takes time for the brain to logically process the incoming data, your mind and your body begin to react almost immediately.

Incoming data passes through two layers of processing. The first of these, the primal layer, is designed to keep humans alive. The brain processes information in this layer quickly; they are reactionary and thus do not include logic (fight-or-flight reactions). In the primal layer, a person begins reacting—whether to words, visual cues, or auditory signs—within five seconds. By contrast, the second of these, the logical layer, takes much longer to process the information—about six times as long (LeSage, Dyar, and Evans 2011). Yes, it can take 30 seconds to process information in the logical layer of the brain.

Unfortunately, the first reaction is usually in defense of a person's current position. Once this primal reaction has occurred, it becomes more difficult to incorporate anything that is learned by the logical layer. When this happens, a person's judgment is clouded by the desire to be right and by emotions such as anger. Being conscious of this phenomenon is important—particularly for anyone who manages people at an emergency scene.

## Discussion Questions 1–4

Have you ever caught yourself reacting to a situation before considering what was actually said or done? Were you able to change your course of action and consider your course of action objectively? What happened, mentally, to cause you to change your course of action?

### Case in Point

I was driving home one morning along a four-lane divided highway. Shortly after passing a slower vehicle in the right lane, a third vehicle sped by honking and yelling at me. Without missing a beat, my middle finger was up and my foot slammed down on the accelerator. About 30 seconds later I realized how poorly I had reacted. I put my middle finger away, took my foot off the gas, and pulled into the right lane. —Craig Nelson

## Jumping off a bridge

Would you follow your crew or a commander's orders if they were going to unnecessarily send you to your death? Probably the first and most obvious answer is no. Would you jump off of a bridge just because your friend did? Even while the verbal answer provided is usually no, in reality, many people do. Numerous people have followed others to their death. For example, in several instances, airline crew members had a hunch that something was wrong or even knew the actions of another crew member could likely lead to a crash but refused to speak up. As ridiculous as it sounds, such crew members jumped off the bridge figuratively—and brought everyone else on the plane with them. Simply asking why would have likely averted a disaster.

This behavior happens for a variety of reasons, with varying degrees of influence depending on the situation. Two common causes of this behavior in the fire service are a desire to develop, strengthen, or maintain crew unity combined with a hierarchical culture of not questioning a senior firefighter or officer. These two causes often work together in the fire service to intensify the need to follow the crowd (or jump off the bridge). This behavior is illustrated by the Abilene paradox, which is where everyone follows the group in one direction but each privately wants something different; they follow the group because it was easier to go along with them than speak up. Most readers may be thinking, "I would never do that." Yet, as far-fetched as this may seem, research and case studies repeatedly support that the behavior prevails.

## Discussion Questions 1–5

When you have jumped off the bridge with your crew, what was it that caused you to do that even though you had other ideas? (It could be as simple as agreeing on a meal or restaurant.) How much effort would it have taken for you to go against the group?

---

### Case in Point

A crew of three is responding to a fire with multiple reports. It is 0200 hours and the crew had been sleeping after a busy day. The engine driver is heading west on 10th Avenue approaching a T intersection, where it is necessary to make a left or right turn decision. The driver begins signaling a right turn. The captain tells the driver to make a left turn, thinking that the right turn the driver wants to make will take them in the wrong direction.

The driver follows the captain's orders and turns left. Unfortunately, the captain was wrong but had his mind on several other things at the time. The crew arrives at the fire without much delay but did not take the most direct route.

During a tailboard talk after the incident, the third firefighter admits that he thought the driver was correct but did not want to say anything. The captain realizes that, although he considers himself fairly easy to get along with, there are dynamics at play that cannot be overcome. The crew was not comfortable questioning the captain's decision to turn left. —Dane Carley

## Getting gray hairs—recognizing stress

Going against the flow, having someone point out that there may be a better way to do something, encountering a scene you have never seen before, or trying to multitask—all of these contribute to an increased level of mental stress. Unfortunately, mental stress reduces a person's ability to make logical decisions. It does not take a Mayday call from a firefighter to raise a person's stress level. In fact, Weick (1990) found that it is the accumulation of smaller stressors that most affects decision making (fig. 1–1). An emergency responder's stress level may start rising for many reasons. As a crew or as individuals trained to recognize stressors, we are capable of making a conscious effort to deal with those stressors. When specific stressors are recognized, the crew should slow down their breathing, slow down their driving, slow down their decision making, and evaluate the entire scene as accurately as possible. Some stressors responders may encounter include:

- A flash hood not donning properly
- The nozzle being left in the wrong position from a prior call
- Accidentally dropping a shouldered hose bundle and making spaghetti at a working fire
- Being unable to intubate a patient who is not breathing
- Responding to an emergency incident involving children
- Encountering inclement weather conditions (snowy, wet, cold, hot, windy)
- Navigating in darkness or a lack of light
- Having poor or no communications (radio problems, lack of CAN [conditions, actions, needs] reports, egos, misinterpreting orders)
- Lacking resources
- Working while ill (with anything from a headache to influenza)

Signs of stress include:

- Increased respirations and shallow breathing
- Feeling overwhelmed

- Tunnel vision
- Moving inappropriately fast or slow
- Poor concentration

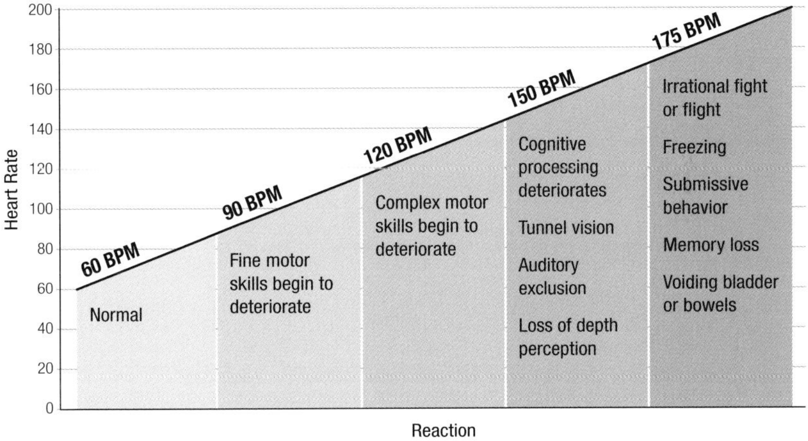


**Fig. 1–1.** Physiologic effects of stress (*Source:* Adapted from Brush 2013, 25)

## Discussion Question 1–6

What other stressors might you encounter besides those that are listed above?

### Case in Point

You are responding to a call involving a child who is not breathing. What is your physical and mental reaction? Your heart rate increases. As this happens, your fine motor skills deteriorate, your ability to hear is worsened, and tunnel vision sets in. Even so, unless you know this will happen, you will probably not notice. The driver will be more likely to speed. The entire crew will be more likely to take risks without properly evaluating the scene. Additionally, a negative outcome is likely to add stress and affect the crew's performance at the next call of the day or the next similar call down the road.

# PUTTING IT ALL TOGETHER

Incident commanders, company officers, and firefighters often find themselves in situations involving all the human factors. Everyone on scene, not just incident commanders, is required to make many decisions quickly in dynamic and complex environments. We do this without having the information or time needed to make an entirely accurate decision 100% of the time. This creates stress that builds from the time of dispatch until the incident reaches the point when it begins improving. We also often attempt to do more than one thing at a time (remember that the brain cannot multitask).

*Running an incident can be like sucking from a fire hose.*

—*Craig Nelson*

As an example, an incident commander managing an incident observes the scene and then predicts future events as best as possible given the available information. This includes assigning companies on the basis of that prediction, tracking incoming resources, listening to the radio, adjusting strategy and tactics as necessary, and most of all, keeping track of all the crews. The process that helps us make this multitude of decisions quickly in the dynamic environments where we work is called recognition-primed decision making (RPDM) (Klein 1998). RPDM, in simple terms, is when the brain quickly scans through memory for a similar situation in the past to provide an answer for the current situation. It can be efficient and effective, but RPDM has a downside. Our mind may choose a similar situation from the past and perform the same actions, but that does not mean that the actions at the past incident were the best options.

Although RPDM is central to how all of this happens, it has a flaw. RPDM influences current decision making but is based on old beliefs or prior practices. Each member on the scene may have a different awareness of the situation because of different education, training, and experiences including past fires, EMS calls, or accidents. It is the fundamental hope that these past experiences will help crew members provide relevant and timely feedback to decision makers. This feedback may cause some uneasiness or may even feel like a challenge to the decision maker's authority. In a department with a learning culture however, crew members should feel comfortable enough to speak up.

Does this mean we must accept that being an emergency responder is a role ripe for mistakes and accidents? Given that we are all human, is it impossible to overcome human factors? No, this means that we use methods to improve emergency repsonders' awareness and abilities to make sound decisions. Some different methods to address human factors might include:

- Using an aid to assist with incident command; this is an extra brain with an extra set of eyes, ears, and hands
- Moving incident command out of a noisy or climatically uncomfortable environment
- Eating healthy and exercising
- Getting sufficient rest
- Stopping or slowing down to understand the situation and make better decisions by increasing awareness, especially when the emergency is not going as predicted after employing initial strategies and tactics
- Taking deep breaths to control your respiration
- Clarifying communications or orders
- Using a command board, paper sheet, or software to track people and resources to reduce the amount of information in your working memory
- Encouraging, listening to, and actually considering input, even at an emergency scene, because the time spent does not really seem that excessive when compared to the improved outcome
- Being mindful of the human factors in yourself and others; develop personal flags, which are human factor trigger points that you recognize as reducing awareness and compromising decision-making ability
- Calling for sufficient resources; fire grows exponentially but we typically call for resources linearly

- Providing rehab facilities and hydration as appropriate given the weather conditions
- Lighting up the scene
- Staying home when you are sick
- Practicing methods or training with equipment with which you are unfamiliar while adding other stressors

## Exercise 1–2

As a crew exercise, put pieces of an SCBA in a windowless room. Enter the room with no lights, loud music blaring, and a time limit. With these three stressors, attempt to assemble the SCBA without cross-threading the connections. For readers without access to an SCBA, attempt to tie a complex knot behind your back or intubate a mannequin in a similarly stressful environment.

## Exercise 1–3

As a command exercise, choose a group of four people. Have three of these people sit next to each other in chairs. The fourth person reads a passage from a book to the person sitting in the middle. With the middle person's permission, the other two seated individuals poke and pat the listener in between them. The two seated on the outside also say random numbers and words. At the end of the passage, the reader asks the listener, who is sitting in the middle of the three, what was said.

The potential list of human factors is never-ending. These are just a few examples to pique your curiosity and help you to recognize how human factors can affect operations negatively but can still be dealt with positively.

### Case in Point: A Tale of Two Fires

A fire occurred in one of the far north regions of the United States in the middle of January. The temperature was below zero degrees Fahrenheit, and the wind was blowing at 20–30 mph. A chain restaurant reported smoke from near the water heater. It quickly grew into a three-alarm fire with many hose lines in operation. It also included a steady operation using street department graders and plows to prevent the water from freezing into a virtual hockey rink on a busy street nearby.

A second fire occurred in the same city on a warm September day. This fire was in a large apartment complex. It quickly grew into an even larger, five-alarm fire involving many separate operations.

A review of both incidents brought forward the point that although the fires were similar in nature (they were even both fought by the same shift of the same department), they were handled very differently. The January fire had a very well-organized command structure with all entities involved present in the command post. The incident command structure in the September fire was not as organized, had communication difficulties with law enforcement, and was never in a set location. As a result, there were several safety issues at the September fire, including a Mayday call.

Why would this be? A warm-weather fire on the surface seems like it would be easier to deal with from an incident command or operations point of view. One reason suggested as to why the January fire went better was that it was too cold to be outside. The person bringing this forward noticed that the chiefs and heads of all entities involved were forced into a command vehicle where they could all work more effectively together in a quiet, warm, and easy-to-communicate environment. A successful outcome is often influenced by human or environmental factors that we often take for granted, rather than the incident commander's abilities.

## SEVEN PEOPLE YOU DON'T WANT TO WORK WITH!

Another aspect of human behavior is recognizing hazardous attitudes. Fire departments are made up of people with different backgrounds, values, beliefs, and life experiences. This in turn contributes to different behaviors: who each of us is, and how we each act all day, every day—they are our human factors. Some may say, "Who cares?" Asking

| Hazardous attitude | Description | Where you might find them | How to prevent or reduce the behavior |
|---|---|---|---|
| Pressing | This is the person who will do whatever it takes to get back to the station. This person tends to do a partial overhaul job and call it good to get back to the station for meal time or rest. | This person may often be an older member who has lost the motivation to do the job right. | This person needs to be motivated/pushed toward the goal of doing the job completely and correctly. |
| Resignation | This person has given up, believing that whatever he or she does, it won't make a difference. This person doesn't contribute to the team. The attitude can be caused by many different situations where the person was teased, ignored, turned away, or even experienced a traumatic incident. | This person can be found at any age or experience level because the attitude is often caused by personal experience(s). Look for people whose ideas and contributions have been rejected without explanation. | This person needs personal guidance/explanation about the importance of his or her role as a member of a crew that works as a team. If a four-person team has only three members who contribute, it is working at only three-fourths of its full capacity. It may help to find the root cause of the resignation and work from that point. |
| Invulnerability | This person believes that he or she is superhuman and can't get hurt. This person takes unnecessary risks. Every event where he or she doesn't get hurt just adds to the ego problem—continually making the attitude more entrenched and the person more dangerous to everyone. | This attitude may likely be found in a younger member of the department because the judgment part of our brain does not fully develop until we reach our mid- to late 20s. Think about the decisions you made in high school. As we write this, we reflect on the miracle that we both survived the risks we took in high school! | This person needs a review of the firefighter line-of-duty deaths and near-miss reporting system to help him or her fully understand the hazards inherent in firefighting. |

| Hazardous attitude | Description | Where you might find them | How to prevent or reduce the behavior |
|---|---|---|---|
| Air show syndrome | This person treats every incident the same by thinking, I've seen and done this 100 times before this way. This person is complacent to the *fact* that every incident/fire is different and must be treated as such (think room-and-contents fires). | This attitude might be found in your 2-20s. Think of the person who might say, "Hold my beer and watch this!" | This person needs a review of the incidents that actually harm and kill firefighters. Many times these are the incidents we consider easy. This person also needs to understand that it is all right to not know every situation he or she will encounter. |

## Discussion Questions 1–7

Have you or any department members displayed any of the hazardous attitudes? (Everyone has most likely exhibited at least one of these traits at some point.) If so, how can they be overcome?

---

The subject of human factors is enormous. This touches on some of the more common ones that affect the emergency services. The key to understanding human factors is to be open to constantly learning about how people react to the environment around them and how their reaction affects that environment.

# REFERENCES

Brush, B. 2013. *The Missing Pieces of Firefighter Survival: Investigating the Psychology and Physiology of Surviving on the Fire Ground*. Emmitsburg, MD: National Fire Academy.

International Association of Fire Chiefs. 2003. *Crew Resource Management: A Positive Change for the Fire Service*, 3rd ed.

Jensen, M. A. 2005. *The Relationship of the Sensation Seeking Personality Motive to Burnout, Injury, and Job Satisfaction among Firefighters*. New Orleans: University of New Orleans.

Klein, G. 1998. *Sources of Power: How People Make Decisions*. Cambridge, MA: MIT Press.
LeSage, P., J. T. Dyar, and B. Evans. 2011. *Crew Resource Management: Principles and Practices*. Sudbury, MA: Jones and Bartlett.
Putnam, T. n.d. Interview by T. Mason, Bureau of Land Management.
Weick, K. E. 1990. "The Vulnerable System: An Analysis of the Tenerife Air Disaster." *Journal of Management* 16 (3): 571–593.

2

# IT ALL STARTS WITH A LEARNING ATTITUDE

## CHAPTER OBJECTIVES

- Describe a learning culture
- Explain the importance of a learning culture
- Discuss methods of learning from near-misses that involved human factors

## THE LEARNING CULTURE TOOLBOX

Building a learning culture in your department is the critical starting point toward improvement—but what is a learning culture? People who work in emergency services attend a lot of training; hopefully, the training never ends. Is this a learning culture? It may be a part, but the simple act of attending training does not define a learning culture. Some emergency responders have attended college or even graduate programs. Is this a learning culture? Again, it may be a part, but this is not a learning culture by itself either.

A learning culture goes further than just attending training or earning a degree. A learning culture is a behavior, it is an attitude, it is a way of thinking and acting. A learning culture means that you and the people in your department seek out information about the emergency services with the purpose of making themselves, their department, and their profession better. A learning culture moves like a wave with momentum (or a snowball, for those who live in the colder climates). As it is cultivated and encouraged over time, it gets easier and eventually becomes habit. Nevertheless, it can slip away and start moving in the wrong direction. As it does, it picks up more steam (or snow) and becomes harder to get going in the right direction again. Changing a culture or creating a stronger learning culture requires that we be open minded. Therefore, we ask our readers to first be open minded and adaptable throughout your entire time in the emergency services—because that is part of a learning culture.

> *It is not the strongest of the species that survives, nor the most intelligent that survives. It is the one that is the most adaptable to change.*
>
> —*Friedman and Mandelbaum* (*2011, 4*)

Chiefs and leaders must encourage their people to speak up and then must listen when their people do. Those working in the street have the mass effect—the wave of momentum—so it is important to speak up, in an effective way, regardless of rank. Each person plays an important role in making sure the momentum (the wave or snowball) moves toward improvement and excellence. The reverse is also true if we do not strive to improve and to seek out better ways of operating: Negativity will not reflect upon leadership alone; it will reflect upon each of us. In today's media-rich environment where everything is recorded on video, we are working under the intense scrutiny of the people we have sworn to protect. The public expects excellent services for their money and accurate answers to their questions. A learning culture ensures that we not only meet but exceed our public's expectations. A learning culture is one in which all people in the department are not merely *taught* to ask why; it is one where everyone *does* ask why.

## THE WHY TOOL

Benjamin Franklin is often credited as having said, "The definition of insanity is doing the same thing over and over and expecting different results." That makes sense—but how do we find new or better ways of responding to emergencies and serving the public so that we do not continue doing the same thing over and over? The answer is, by asking why. We all need to ask this, because the behavior we are looking for is curiosity. When we are curious, we have the courage to ask why, we work to seek answers, and we enter a learning culture. If, however, we are discouraged from speaking up because we are intimidated, afraid, new, young, or of lower rank, then we are not in a learning culture. If we are discouraging a learning culture or our people are not speaking up and asking why, then we are not striving to improve, and this leads to a stagnant department. Working in the emergency services should instill the values of pride and excellence, and asking why is the tool we use to start the process of striving to improve.

The why tool can be implemented in many ways, such as:

- "Why did we do it that way?"
- "Why didn't we do it that way?"
- "Why do we respond to these calls?"
- "Why did we use that tactic?"
- "Why do we clean the station every day?"

These questions can initially seem offensive, as if the intent is to second-guess the current operation or make a person look stupid. However, all they really seek is truth and improvement. Once people within a department become accustomed to hearing why, the question quickly loses the offensive feel and begins to sound normal. It takes time and practice to encourage other people to ask why. It also takes time and practice for you to learn to respond without feeling threatened when you are asked why. There is usually a positive intent behind the initial question, but we often forget to ask the person why they asked in the first place. Too many times when people get up the courage to ask why, it is quickly stifled by someone responding, "Because we have always done it that way." This implies that either there is no good reason for it or the person answering does not know the real answer. This defensive response also sounds negative, effectively discouraging anyone within earshot from asking why ever again.

On the other hand, the answer to a person in a learning culture who asks why would be similar to, "We do it this way because this is safer, more reliable, more effective, more efficient, or cheaper." The point is that there needs to be a reason. Moreover, a reason supported by research or data (or both) carries the most weight because it is an objective, measured response. This is where learning begins for both individuals and departments—by simply asking why.

## Case in Point: The Parable of the Monkeys (adapted by Ryan Tracey)

### *The experiment*

In a room, there were five monkeys. A bunch of bananas hung from the ceiling, and a ladder stood nearby. Inquisitive and hungry, one of the monkeys climbed the ladder and reached for a banana. As soon as he did so, a scientist opened a porthole and drenched all the monkeys with a high-pressure hose.

After a few minutes, angry and dripping wet but no less inquisitive and hungry, another monkey decided to give it a go. He climbed the ladder and reached for a banana. As soon as he did, the scientist opened the porthole again and drenched all the monkeys with the hose.

After a few more minutes, another monkey decided to give it a go. This time his mates would have none of it. As soon as he touched the ladder, they rallied together and beat him up.

None of the monkeys dared go near the ladder any more.

The next day, the scientist removed one of the monkeys and replaced him with a new one. Because this monkey was not aware of the consequences, he headed straight for the ladder. The other monkeys headed him off and beat him up.

The next day, the scientist removed another monkey and replaced him with a new one. This monkey, also unaware of the consequences, headed straight for the ladder. Again, the other monkeys headed him off and beat him up.

This continued for a few more days. Each time, like clockwork, the new monkey would head for the ladder, and the other monkeys would beat him up.

Then on day 6, a strange thing happened. Yes, the new monkey headed for the ladder, and the other monkeys beat him up. However, none of those monkeys was an original from day 1. They all dutifully beat up the new guy, but they had no idea why!

### *Monkey business*

I'm sure we've all had times when we've felt like one of those monkeys. We do something a particular way *because that's the way we've always done it.*

## Discussion Questions 2–1

Do people in your department ask why? If so, what does their intent seem to be? If someone asks why, what is the reaction?

Stefan Svensson says the Norwegian Fire Academy has trained their new firefighters to ask why from the start of their fire service career; the teachers encourage this question. What are firefighters taught in your department's training? Have your firefighters had the unproductive experience of hearing, "You have two ears and one mouth. That means you should be listening twice as much as you are talking!"

---

# THE HOW TOOL

After we ask why, we should be asking how. This is the natural progression of growth in a learning culture that pushes the snowball in the right direction. The how tool can be formed in many ways, such as:

- "How can we improve?"
- "How can we provide that service?"
- "How can we improve our operations or training?"
- "How do we get over that hurdle?"
- "How do we fund this?"

The how tool is where many get stuck, though. Have you ever heard either of these phrases: "Why would we want to do that?" or, "How would that ever work here?" These types of statements kill personal, departmental, and professional improvement. The Wright brothers would have never invented one of the most important machines in existence, the airplane, had they listened to everyone else around them. Had Thomas Edison given up after the first thousand tries to invent a better lightbulb, we might still be using candles and oil. Thomas Edison's response to his many failures was, "I have learned many ways that a light bulb won't work." We can go on and on (Louis Pasteur, Henry Ford, Galileo) but the point is that all these innovators brought about new and better ways of doing things, even while their ideas were thought at the time to be completely wrong or impossible.

To get past the first step, we ask why and then follow it with how. The snowball is now rolling in the right direction because you are asking why and how. However, the people we need asking those questions are human. Therefore, understanding human factors and how they influence people is a key part of developing a learning culture.

### Discussion Question 2–2

Why is the how tool important?

---

## WHOA! THIS LOOKS LIKE MATH—I'M OUT!

The vast majority of near-misses, incidents, and accidents can be traced back to one or more of the human factors discussed in chapter 1. If we continue to learn (ask why) about human factors and understand how they affect us, our work becomes much safer, and we become more effective as individuals, departments, and a profession. We need to collect relevant data on how human factors influence emergency services. Why? Because we need to know what is really causing our safety problems. The data collected can be used to identify positive and negative trends within specific areas of interest. Identifying trends allows us to develop training to overcome, accommodate, or regulate behaviors to maintain or improve performance of the department and the profession. Looking at firefighter near-miss statistics (fig. 2–1) shows that human factors contribute to a strong majority of incidents (almost all of them).

### Discussion Questions 2–3

How many of the categories in figure 2–1 influence or are influenced by human factors? Does this change your perspective of the cause of incidents, accidents, and near-misses?

---

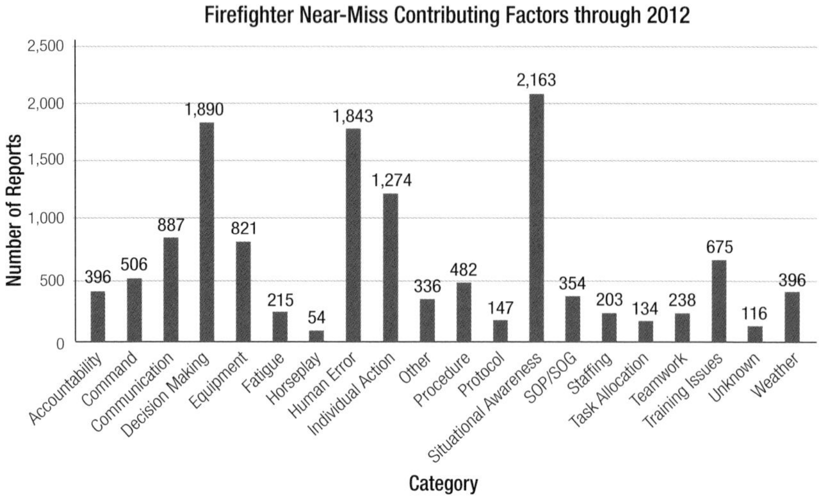


**Fig. 2–1.** Number of times various factors contributed to a near-miss incident (*Source:* National Fire Fighter Near-Miss Reporting System)

Based on the categories presented in the firefighter near-miss statistics in figure 2–1, all of the categories except for "equipment" and "other" can be directly related to human factors; there are even good arguments to include those categories in some cases. Have you ever considered how these factors affect you, your crew, your equipment, and your daily operations? These 16 root causes lead to most of the incidents in our industry, yet we focus on a multitude of symptoms of these root causes. What symptoms are we talking about? Here are a few examples:

- Technological answers to members not wearing their seat belts (safety lights, safety bells, red seat belts)
- Improved turnout gear, rather than more fire behavior training
- Ever-thickening rule books, heavy on standard operating guidelines and procedures (SOGs and SOPs)

We spend the bulk of our time in the technical fix area of diminishing returns while ignoring the area of human factors, which is where we have the most potential for individual, departmental, and professional performance enhancement.

### Case in Point

A young firefighter is responding to a traffic accident on the freeway during rush-hour traffic. He is riding a forward-facing jump seat. It is a warm evening, and he is enjoying the ride. As the medic engine enters the cloverleaf on-ramp, he reaches under his seat to retrieve a box light since it is starting to turn dark. As he sits back up, the sleeve of his turnout coat catches the door handle and opens the door. The door swings open because of the force of being on the outside of the turn. He begins to grab for something to hold onto but can't find a handhold. He sees his uniform boots that were sitting on the floor of the engine after donning turnout gear fly out into the grass. Luckily, he is wearing his seat belt. It is only a lap belt on this older engine, but it keeps him in the engine. He closes the door after the engine leaves the cloverleaf once he can overcome the force.

Fast-forward five, ten, or even fifteen years. This same firefighter is riding to a variety of calls. Sometimes he wears his seat belt, and other times he doesn't. Maybe the rationale is that the call is "just around the corner," or "I need to put on my SCBA," or "I trust this driver to be safe"—the firefighter justifies not wearing his seat belt many different ways. Regardless, it comes down to one area: human factors. Specifically, this is an example of the hazardous attitude of machismo. –Dane Carley

Firefighters have known for years that they should or must wear safety belts. There have been many injuries and deaths over the years because firefighters were not wearing their seat belts. Many improvements have been made to the fire apparatus and seat belts. We have enclosed cabs, added three-point harnesses, and put in lights and bells indicating when a safety belt has not been engaged. We even have SOGs, rules, signs, and national pledges that tell us to wear our safety belts. So, why do we still see the fire service behavior of not wearing a seat belt? It is because we have been treating the symptom with technical answers like bells instead of directly addressing the problem, which is rooted in behavior. The problem is human behavior; therefore, the solution lies in human factors.

Until those of us who work in the fire service begin to focus our learning on human factor–related data, we will continue to restrict our capabilities. Part of what makes a profession stand out is that they collect and analyze data to determine trends. Then, they strive to follow through to increase performance.

Figure 2–2 illustrates the proportion of fire service incidents caused by machines as compared to humans. Given the amount of information we now have available, we are generally paying too little attention to human factors. Now, when we honestly look at the root causes and acknowledge that they are largely related to human factors, we can begin to implement more effective methods for improving the human performance part of our operations.

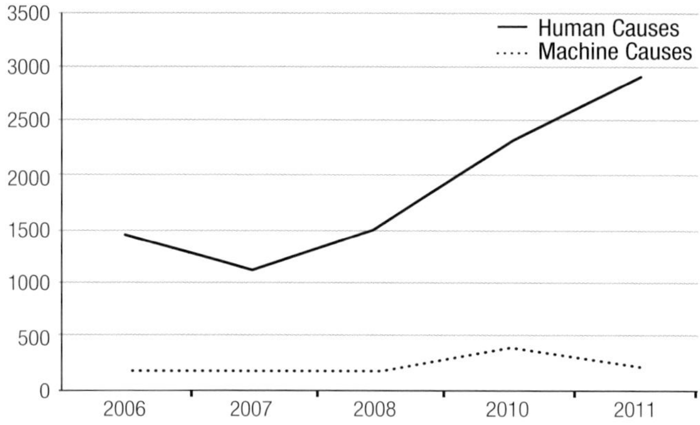


**Fig. 2–2.** Proportion of fire service incidents caused by humans compared to machines between 2006 and 2011, except for 2009, for which no data was available (*Source:* National Fire Fighter Near-Miss Reporting System)

## Discussion Question 2–4

What percentage of your department's training time is spent on human factors as opposed to technical training and skills?

Many reasons may be given for not accepting this approach such as resistance to change, fear of the unknown, the work involved with the new methods, or tradition. Although these are understandable, they are not valid reasons; they are human factors playing a role in decision making.

A department striving for improvement operates in a persistent learning culture because learning is a fundamental component of improvement. Therefore, departments striving for constant improvement not only collect and analyze data but also perform root-cause analyses of incidents. These analyses typically find that humans' reactions to the environment in which they were operating contributed to the incident. However, the focus is not on *who* made the errors but on *why* the environment in which the person operated led to the errors. In other words, rather than place blame or ridicule an individual, the process is designed to provide an opportunity for the entire department or profession to learn from one error that has already taken place, instead learning it one by one.

Root-cause analyses focused on human factors often find that department culture, stories, and structure lead to the pattern of decision making contributing to the near-miss rather than an error made by a single individual (LeSage, Dyar, and Evans 2011; Carley 2010). Once a department looks at their operations through this lens, they can adjust their operating environment. These adjustments range from engineering controls to organizational behavior. The goal is to prevent future incidents stemming from the same root cause as past incidents.

## Real-Life Example

### Event description

Our engine was dispatched to a reported structure fire at a local apartment complex, with the callers reporting visible smoke and fire. Upon arrival, we found a three story apartment building in the complex showing heavy fire from the A/B corner, extending from the first division up and showing through the roof. My initial assignment was to throw a 24-foot extension ladder, which was quickly changed to operating the engine's deluge gun to achieve a quick knock down and allow us to enter the affected apartments.

I climbed up on the truck and disconnected the deluge gun from its base, in order to place an extension (pipe) on it to allow for greater reach and penetration. Placing the pipe was a standard practice among our shift's crew, and when it had been done in the past the engineer would ask the firefighter if he was ready for water prior to charging the deluge gun. On that shift, the engineer was acting captain and a firefighter from another station was detailed to drive. Shortly after I had the deluge gun disconnected from its base, a jet of water came up and knocked me

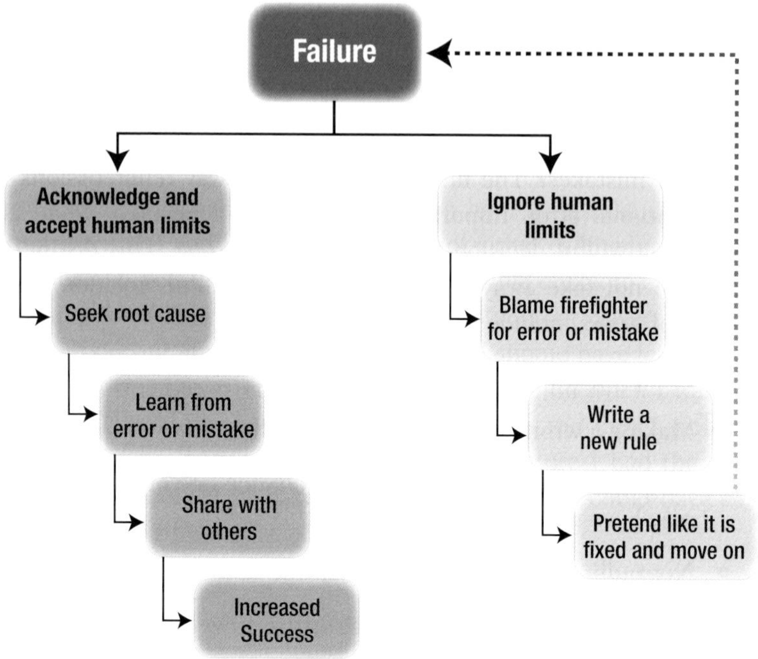


**Fig. 2–3.** Two ways to handle a failure. The method at left is not used often enough in emergency services because the tendency is to use the method at right.

This is a subtle but critically important distinction. The first mistake we often make when critiquing someone else's actions is not placing ourselves in the decision maker's shoes. To understand the decision, we must consider what the decision maker knew when the decision was made instead of what is known after the event was over. The second mistake is looking at the faults of the decision maker. What was his or her training? What was his or her certification level? What was his or her ______ (fill in the blank)? The questions should go further. If the answer is that the decision maker had insufficient training, then the next question should be why.

## Case in Point

A wildland crew boss takes his crew from a safe location toward a second safe location to engage a fire. This decision is based on a weather forecast, experience, and distance to the second safe point. The crew boss waited for the predicted wind shift and then began moving his crew. What

> the crew boss did not know was that there was a second wind shift coming his direction. The crew was caught between safe zones in an inescapable situation.
>
> This is the same decision any competent crew boss would have made given the exact same situation based on what the crew boss knew at the time. The environment around his crew, which was beyond his control, changed. Does this mean he is a bad crew boss? Does this mean he made a poor decision? No, it just means that he is human. He made the best decision possible with the information he had at the time.

Not punishing unintentional errors does not imply that people are no longer to be held accountable. Also, making the same mistake more than once is no longer an error because (after retraining is offered) it implies a conscious decision to act that way. People are held accountable through methods more effective than punishment. The overall goal of these methods is to increase performance. This can be accomplished only by creating a trusting environment for learning. There are basic steps to accomplish this. First, there is an expectation that someone making an error will report it during an informal tailboard talk, engage in formal critique, and submit it to the near-miss reporting system for further discussion. Second, there is an expectation that the department's personnel will receive training to avoid the error and improve performance in the future. Third, the department will evaluate itself to ensure department practices or culture are not preventing the adoption of best practices or widely accepted skills and methods.

People make the decisions they do because of their training, experiences, education, and department culture. Identifying a cultural bias, or even a fire service bias, is easiest with valid and reliable data, which help a department identify trends that are not limited to a single incident, crew, or shift. Analysis of the data reveals trends (good or bad) that pervade the department, often originating in the department's culture or training division. The point is to use the data to reinforce the positive trends and retrain in light of the negative trends. However, data can be collected only if the environment encourages and allows it.

Will there still be pressure to hold someone accountable? Most likely, the answer to that question will be yes. Imagine holding someone accountable for an honest error by punishing them. Has it solved the problem? Did it prevent it from happening again in the future? Did anyone learn from it? No, no, and no. Problematically, holding someone accountable for such an error through discipline will cause that person—and everyone else in the future—to avoid reporting errors that are made. Data cannot be collected to improve performance because people will not report errors that will embarrass them or get them in trouble.

## Discussion Question 2–6

When you have been punished, what did you learn from it, and what did others learn from it?

---

### Case in Point

A young firefighter is out on a grass fire in brand-new, shiny yellow brush gear. It is not much of a fire, only a few acres of grass, and it is contained. The firefighter finds a hot spot smoldering in some logs. He decides "blackening" his helmet in the hot spot and rubbing his gear in the soot is a good idea.

The chief catches wind of the incident. His reaction is to call the young firefighter on the carpet for his actions. Were the firefighter's actions intentional? Yes. Were the firefighter's actions detrimental to his safety by degrading his personal protective equipment (PPE)? Yes. Was it an error or a violation? An argument could be made either way in this particular case, so the chief's actions fit the expected response for a willful violation.

For a second, consider another side of this story. Ask yourself and find an honest answer to "Why did the firefighter do this?" This firefighter was trying to fit in with his senior firefighters; he was trying to be "one of the guys." Cultural expectations pushed this firefighter to make a poor decision. So, how else could the chief have handled this situation? It is obvious that holding the firefighter accountable is necessary. Still, the chief alienated this firefighter by disciplining him because the firefighter actually felt he was making a valid choice. It may have been more effective for all involved to hold the firefighter accountable by having him teach a class on PPE (as one example). This would have opened the firefighter's eyes to the dangers of his actions and possibly started a cultural shift away from department members' using dirty gear as a measure of a firefighter's worth.

# REFERENCES

Carley, D. A. 2010. *Organizational and Group Influences on an Individual in a Fire Department Company*. Fargo, ND: Fargo Fire Department.

Friedman, T. L., and M. Mandelbaum. 2011. *That Used to Be Us*. New York: Picador.

Klein, G. 1998. *Sources of Power: How People Make Decisions*. Cambridge, MA: MIT Press.

LeSage, P., J. T. Dyar, and B. Evans. 2011. *Crew Resource Management: Principles and Practices*. Sudbury, MA: Jones and Bartlett.

National Fire Fighter Near-Miss Reporting System website. http://www.firefighternearmiss.com/.

Tracey, Ryan. n.d. "The Parable of the Monkeys." http://ryan2point0.wordpress.com/2011/03/08/the-parable-of-the-monkeys/.

3

# MANAGEMENT IS NOT LEADERSHIP, AND LEADERSHIP IS NOT MANAGEMENT

## CHAPTER OBJECTIVES

- Explain the roles of a manager and a leader in a department by illustrating the differences between the two
- Identify key leadership traits
- Describe key leadership behaviors
- Present methods of improving personal leadership skills and abilities
- Propose ways for a department to improve processes for choosing people with leadership skills and abilities

A learning culture does not develop in a vacuum. Learning is the foundation of any department striving to improve. However, a learning environment must be developed in a supportive environment—namely, one enriched by leadership. This chapter provides a basic understanding of what the leadership environment is, how to identify leaders, and how to apply leadership behaviors to daily activities. First, though, it is important to understand what distinguishes leadership from management.

Leadership requires encouraging constructive conflict in an effort to improve ourselves, our departments, and our profession because constructive conflict is the source of innovation. Good leaders encourage different opinions and then listen to them even when they do not agree. This is how groundbreaking and innovative methods, practices, and tactics emerge. Leaders are able to restrain their emotions until they can cognitively process and weigh the information presented on the basis of merit and then apply the ideas and feedback that will improve the department and the profession. Good leaders within the ranks bring forward ideas and feedback to good leaders in administrative positions. Departments often put people through a gauntlet to hire the best but then drop the ball when it comes to listening to them just because they are new.

Leadership involves being secure with yourself and your position. Leaders are not threatened or intimidated by others around them who push change, have superior technical skills, are highly intelligent, or are well educated. Instead of fearing that those around them may be smarter, may make them look stupid, or may take their job, leaders encourage, push, and support those around them to become better and succeed. This approach gains the power of the many to move the department forward. This is how big things get done. The power of a collective group far outweighs that of an individual. Group success brings accolades and positive reinforcement for both the group members and the leader. The leader in turn is successful because of the success of the organization—not because they were the smartest or most talented, but because they recognized the talents of others and used them to the advantage of the group. Good leaders seek these people out and work with them to make the department better. They recognize that one person cannot and will not ever know everything and realize that they must constantly strive to learn more and refresh current skills.

Leadership means working harder than everyone else, including those over whom you have authority. Leading is about setting an example for those around you. It is also about setting a standard. Leaders do not expect others to meet a standard they do not meet themselves. People want a leader who leads them, not someone who sits at the back barking orders.

> *Leading is not always fun or easy, but when success is achieved while leading it is very satisfying.*
>
> —*Craig Nelson*

Leadership is about improving yourself and those around you. This means being an avid reader and encouraging those around you to read. Mike Myatt (2010) has observed that most Americans read only one book a year at best, whereas successful CEOs read four to five books a month and the world's historical leaders read as many as one book per day. A focus on improvement means making training and continuing education priorities. Leaders are willing to set aside management objectives in exchange for time to develop themselves and those around them.

> *If you call yourself a leader but are not an avid reader, you're likely not a leader.*
>
> —*Adapted from Myatt (2010)*

Most of all, a leader is a person who is self-aware. Leadership is about knowing personal shortcomings and strengths and then understanding how these affect interpersonal relationships. A leader reflects on less-than-perfect interactions and finds ways to improve in the future. Jack Zenger and Joseph Folkman (2009) have found that poor leaders had at least one of ten leadership derailers. Leadership derailers are specific traits identified by sociologists that negatively affect a leader's potential. Peers and subordinates rated these poor leaders as having obvious flaws that most people would want to fix. Yet, the worse the peers rated the poor leaders, the higher the leaders rated themselves. A good leader spends a lot of time reflecting on interactions and decisions and asking for honest input from third parties to avoid these leadership derailers. Consider knowledge of your own strengths and weaknesses to be your own personal accreditation. A leader tries to be adaptable, open to new experiences, and agreeable but recognizes that no one is perfect.

## Exercise and Discussion Questions 3–2

1. List three additional traits, behaviors, or daily activities that are part of a leadership style (also see fig. 3–1):

   –

   –

   –

2. List three of your own strengths:

   –

   –

   –

3. List three of your own weaknesses:

   –

   –

   –

4. Reflect on your own experiences. How would you describe a leader now? Write a sentence or two describing what leadership is:

   –

5. Do any of your definitions include anything related to title, position, or rank?

---

# LEADERSHIP MYTHS AND MEASURES

Certain leadership traits and behaviors have been identified by social scientists. These traits and behaviors are consistently stronger in leaders (Hughes, Ginnett, and Curphy 2006). If we know which behaviors and traits are most important, then we can measure them; if leadership is measurable, then it is also teachable; and if it is teachable, then it is learnable. Thus, we can teach people to be leaders. Even though a select few charismatic leaders may be born that way, everyone else is still capable of learning leadership.

> *If something is measurable, then it can be taught. Leadership is measurable. Therefore, leadership can be taught.*
>
> —*Dane Carley*

Several leadership assessments exist. Some measure personality traits that are correlated to leadership ability. For example, the Five-Factor Model establishes the Big Five traits, featuring agreeableness and openness to new experiences alongside extraversion, conscientiousness, and neuroticism. The Five-Factor Model has led to the Revised NEO (Neuroticism-Extraversion-Openness) Personality Inventory. Other leadership assessments measure behaviors correlated to leadership ability. For example, the Emotional Quotient Inventory (EQ-i) measures a person's emotional intelligence in terms of a number of behaviors.

Assessments serve different purposes and should be used accordingly. Some assessments measure traits, such as the Myers-Briggs Type Indicator (MBTI), to help a person understand their personal tendencies. It is then up to that individual to determine how these traits cause others to see them and how they affect their interactions. These are typically used to spur self-improvement, especially in the context of team building. Other assessments directly measure leadership behaviors and, in some cases, leadership derailers. These assessments can be used for self-improvement and for promotional processes. Some examples of assessments that can be correlated to leadership traits, abilities, or styles, depending on their application, include:

- EQ-i 2.0
- MBTI
- Minnesota Multiphasic Personality Inventory (MMPI)
- Hogan Personality Inventory (HPI)

It is important to understand how one's self is perceived because other people often perceive us very differently than we perceive ourselves. A leader may think he or she is approachable, fair, and motivational, but the reality could be that no one else sees the leader that way. Importantly, how others perceive a leader determines that leader's effectiveness.

## EFFECTIVE LEADERSHIP IS IMPORTANT

What percentage of a responder's time is spent at an emergency scene? 50%? 25%? In the fire service, it is typical for a responder to spend less than 5% of their time at an emergency scene (Carley 2008). Conversely, 95% of a responder's time is spent working in a situation where time compression is not a factor, so development of the leader's followers—the skills and abilities of those the leader is helping to become better—should be a primary objective. Yet, few departments actually assess their people's leadership ability because doing so is difficult and complex. Thus, many of the testing processes instead focus on the technical abilities needed 5% of the time in an effort to provide an objective process. Just because this is easier to do does not mean it achieves the best results. When it comes to important matters like hiring and promotions that will affect your department for years into the future, doing it in the quickest and easiest way in the short term may be a poor way to operate for the long term.

### Case in Point

A young person joins a fire department. He is excited about the new job and is highly motivated to excel. He studies each night for the first couple of years. After the first couple of years, he begins to watch more television, complain about department operations, the other shifts, and so on. Over time, this person is promoted through the department to become one of the chiefs he complained about not too many years earlier. Now, he is complaining that the firefighters just do not get the job anymore. They do not study like he did, and they do not value the meaning of it. Ironically, the same was said about his generation by the chiefs when he was a firefighter. Why does this happen over and over in the fire service? Why do we continue to see motivated people come into the fire service only to become like all the rest?

This firefighter's department uses a pass/fail multiple-choice test to begin the testing process for captains. The test is based on several textbooks from a reading list that the department provides to the candidates ahead of time. Knowing that, what is this department testing? This department is testing rote memorization. Rote memorization of what? Memorization of a bunch of textbooks that may or may not be directly related to the job. No one knows because the questions from the test bank have likely not been correlated to the department's job analysis. Also, such tests do not typically measure leadership potential.

Nevertheless, the first step is passing this test. Those candidates with positive attitudes, openness to new ideas, the ability to adapt, and agreeableness may miss a passing score by one point. The person with the worst attitude who hates change is one of the few who pass the test. The department defends the process by saying that they will eliminate those with a poor attitude in an interview or during probation. However, the pool from which to choose has already been biased toward those with the ability to memorize without any value given to those with leadership skills. Plus, interviews have been consistently shown to be one of the worst measures of a person's ability to perform a job (Edwards 2005, 97, 104); it has even been said that only psychopaths and narcissists excel in interviews.

Reflecting on the typical emergency services promotional process, it becomes more apparent why departments may have a difficult time growing, changing, adapting, and succeeding. We know why. Our promotional processes are not testing for the knowledge, skills, and abilities that represent the position of an officer (leader). Again, Benjamin Franklin sums it up best: "Insanity is doing the same thing over and over and expecting different results." So why should we expect different results from our personnel if we do not adjust the department's processes?

Technical knowledge is critical to safe operations. A department has an acute need for technically competent people who make good decisions at an emergency scene. However, leadership is critical to a department's long-term success. The need for leadership is chronic because a poor supervisor affects so many people. Captains set the example for upcoming firefighters; battalion chiefs, for upcoming captains and firefighters; and chiefs, for everyone in the department. Most emergency responders develop within the same department because of low turnover, so the only exposure they have to management and leadership styles may be their predecessors. Therefore, a person who learns to manage or lead from a chief as a new firefighter will likely manage or lead similarly many years later, assuming no other outside influences. Consequently, good and bad leaders can affect a department for years.

Leadership is also the key to fostering a learning culture. If we think of the learning culture as a building's foundation, then leadership is the environment in which the building exists.

## APPLYING LEADERSHIP TO DEPARTMENT OPERATIONS

We know what traits and behaviors a leader possesses because of sociology research. We also know that leadership is important to any department—especially to a department striving for excellence. Still, how do we apply this to a department's daily operations?

### Assessing for leadership

Every time a department promotes someone, they are placing that person in a formal position of influence. This should never be taken lightly because the impact that person has could last for decades. If the person has a negative attitude, it will affect many people negatively over the years. If one person affects three people, those three people will go on to affect nine people and so on. However, the opposite is also true of good leaders. When this is framed in terms of a snowball effect in either a desired or undesired direction, the cumulative effect of this cannot be understated. A person who automatically resists any and all changes will, over many years, severely impede a department's progress.

Therefore, one place to start applying leadership theories to daily operations is in hiring and promotional processes. Hughes, Ginnett, and Curphy (2006) have attributed high levels of incompetence among managers to "poorly designed leadership programs, ill-defined expectations, and invalid selection and succession processes" (170). Hogan (2007) has found very similar results. If invalid selection processes are one of the root causes of poor leadership, we should start by looking for people with these traits and abilities to hire and promote. A leadership assessment measures leadership traits or behaviors with the same reliability and validity as a written test measures rote memorization of book knowledge or an assessment center measures technical abilities. The typical civil service promotional process used today measures the same dimension of a candidate in many different ways. Whether a process requires certain certifications to apply, a written test, an assessment center, or any combination of these, it still measures only the technical knowledge, skills, and abilities of the candidate. This is an example of our heavy reliance on management because we are typically testing for and selecting those with management traits through technical written tests and assessment centers. Designing a system that gives leadership abilities comparable weight to technical skills and abilities will be the first step in setting up the department for long-term success.

Subordinates and supervisors alike expect a leader to be capable of teaching others technical skills. Promoting technically competent leaders satisfies that expectation. Subordinates expect their supervisors to be technically competent but also capable of helping them develop into the department's future leaders. Leaders are expected to develop their subordinates using leadership abilities for the department's succession plan while not making technical errors that put the department in a poor public spotlight. Because a department needs to meet this balance, both dimensions are important.

An assessment such as the EQ-i could be added to balance written knowledge with leadership abilities in a promotional process. Knowledge is necessary to accomplish tasks, but leadership is necessary to accomplish everything else an officer does.

## Exercise and Discussion Questions 3–3

1. List in order of importance all of the qualities you believe are valuable in leaders and officers:

   –

   –

   –

   –

   –

   –

2. List the traits your department currently trains and tests for:

   –

   –

   –

   –

   –

   –

Note that if you use typical technical skills–based written tests, then you are often testing rote memory, basic math, reading comprehension, and science abilities. Also note that if you use an assessment center, it may be testing either leadership or management abilities. Finally, note that if you use an interview process, then you are testing speaking and communication skills, past experience, and thought processes.

Do any or all of these actually measure the skills and abilities for which you are looking? Compare the two lists above and then ask yourself, are you actually testing, training, and seeking out candidates for hire and promotion based on what you desire in a leader?

---

## Group development

People identify with many different groups at the same time. For example, firefighters identify with their crew, shift, department, and profession (Carley 2009). Depending on the situation, the strength of that identity contributes to some of their decision making. A stronger identity with the profession may cause firefighters to aspire for goals that cause the crew to make fun of them, or vice versa.

### Case in Point

A crew has a long history of drinking a couple of beers at the station, and even though the department's rules do not prevent it, the public would come down hard and fast if they knew that the crew did this. Some of the crew members feel like this is a great way to build crew camaraderie, despite knowing it could also make the department look bad. Other

> crew members fear that the practice makes the fire service, as a whole, look unprofessional, although they acknowledge the point about building camaraderie.
>
> These two groups of crew members are identifying with different levels of the fire service in this example. The first has a stronger identification with both the crew and the department, but with the stronger tie being to the crew; this overwhelming tie to the crew supports their decision to drink at the risk of tarnishing the department or profession in the public's eyes. The second group has a stronger identification with the profession but does not entirely separate itself from the crew; however, the overwhelming connection to the profession prevails on them to avoid drinking to maintain a professional image in the eyes of the public.

Regardless of which group a person identifies with, there are certain processes that each passes through and certain roles that need to be filled, regardless of the group's size. For example, every group has a person who brings aptly timed lightheartedness or a person who enjoys looking up information and bringing the technical or factual aspect to the conversation. Depending on the group, some roles can be filled more readily than others, based on the group members' abilities and natural preferences.

Deciding who fills these roles is determined in part by the four stages of group development that every group passes through: forming, storming, norming, and performing (table 3–1). In addition to the assignment of each member's role, this process determines the group's long-term values and the professional level with which the group identifies strongest. The reason that storming is part of the process is that members are looking for their role in the group, initially with tension—both inward, at one's self, and outward, at other group members. The members feel tension inward as they try to align their personal values with the group's values, while they feel tension outward as they fight for their spot in the group. Group norms—the group's values—begin to emerge toward the end of the storming phase as they pass into the norming phase where each member's personal values has shaped the group's values. Leaders should begin to actively instill positive group norms early and then reinforce them during the norming phase because this is where the group's values are molded with the most ease.

**Table 3–1.** The four stages of group development

| Stage | Characteristics |
|---|---|
| Forming | Strong dependency on leader with little independent thought; low productivity; group role ambiguity |
| Storming | Determining who fills what role; fitting individual values into the group's values; increased conflict; beginning to establish group norms and expectations |
| Norming | Increasing conformity with the emerging and strengthening norms and values; increased trust and structure as members accept their roles; development of group abilities; increased performance and constructive communication (as long as leaders instilled those values) |
| Performing | Not all groups reach this level; members understand and are comfortable with their roles; comfortable discussing conflicting views; focus their energy on the group's goals; excellent performance |

## Exercise and Discussion Questions 3–4

- What stage is your crew or group at? Why would you say they are at that stage?
- How long did it take to get to the current stage?
- Can you think of other groups that you have been a part of at the other stages?

Unfortunately, for group development, this process starts over each time a new person is added to the group. Crews or shifts that see constant flux do not ever have the chance to pass through the group formation stages and become a strong-performing team. Instead, much time is spent in the first three stages of the group's formation if the changes occur often enough. The graph in figure 3–3 illustrates a fictional comparison between three companies—one static, another in constant flux, and a third with minimal change.

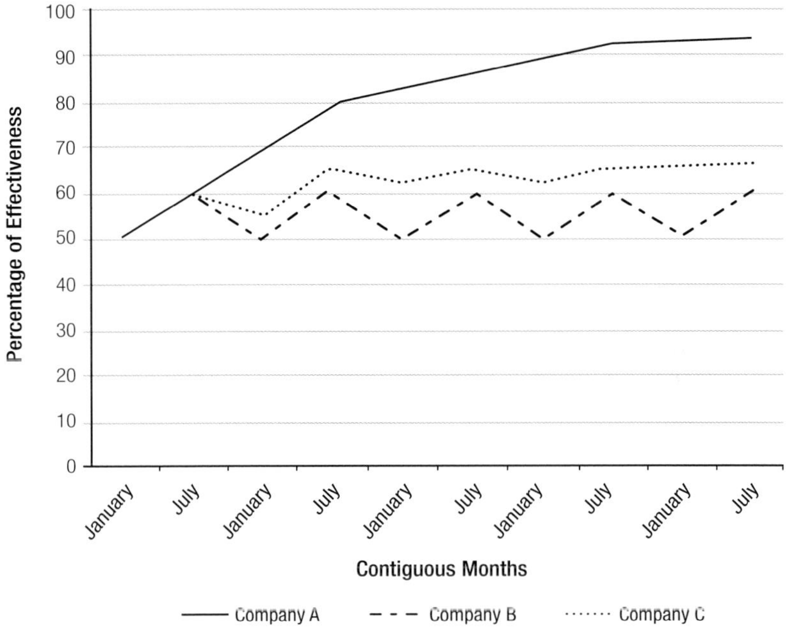


**Fig. 3–3.** Fictional comparison of three companies. The crew members of Company A never change; all of the crew members of Company B change regularly; and only one of the crew members of Company C changes regularly.

Just as Rome was not built in a day, groups do not develop overnight, in a week, or even in a month. It takes six to twelve months for a group to develop. This is important because many try to rush this and then get frustrated when it does not work. Group development cannot be rushed. Group development and maintenance is fostered by (Carley, 2009):

- Developing a sense of belonging to the group (through group identifiers, mascots, mottoes, etc.)
- Conveying a reasonable sense of competition with other similar groups (healthy competition between companies)
- Performing at a high level through training and education
- Clearly communicating the leader's expectations of the group
- Developing goals as a group
- Aligning the group's goals with the department's goals
- Promoting a good work ethic

- Fostering a feeling of empowerment among group members

Contrary to popular belief, just achieving cohesiveness does not make a high-performance group. In fact, the opposite is true over the time during which a group develops. A group that is developed by its leader into a high-performance group eventually becomes cohesive. The points above help a group become a high-performance group, which then leads to stronger group development and longer-term cohesiveness.

A high-performing group becomes cohesive largely because it wants to remain high performing. However, the group's energy should be focused on performing a good job and maintaining a good work ethic because task orientation and pride in performance improve decision making (Mullen et al. 1994). This cohesiveness strengthens the group's performance because it encourages constructive criticism through honest communication built upon trust. Once a crew is recognized by others as performing above the norm, the members recognize that constructive, critical communication is the only way to continue functioning at that level, which builds trust among those members.

Building trust during group development and maintaining it are critical for a group's success. For leaders to be effective, they must build trust among those with whom they work. Trust can be built in many ways, but it can also be broken in many ways. Even though it takes time to build trust, it takes only one action to break that trust, and sometimes the leader may not even know why the trust was broken because it depends on a follower's individual perception of what is right and wrong (Carley 2009). Good leaders know methods for building trust and understand group (crew) development. Here are some ways to build group trust:

- Most of all, listen—actually listen—to learn what your follower's expectations of you are
- Spend time together
- Foster group values
- Encourage a trusting environment that values constructive conflict
- Get to know each other (ask questions)
- Praise individuals publicly

Good leaders also recognize how trust is undermined or destroyed. One of the most prevalent ways to destroy trust is through hazing. Hazing has played a large role in the initiation of many new people in the emergency services. The intent of hazing is ostensibly to build loyalty to a department, station, or crew. Much of the time it is not intended to have a negative impact, but it does. This is another one of those times when we need to step back and ask, "Why are we doing this?" Often the answer may be because it was done to us. As already stated, answering "Because we have always done it this way" should trigger an alarm to go off in your head telling you this practice needs to be reevaluated. Sure, hazing builds loyalty, but it decimates trust.

Here are some ways to destroy trust:

- Hazing
- Mean-spirited teasing or ridicule
- Public punishment
- Saying one thing and doing another (double standard)

## Leadership communication

One basic behavior builds group development and trust: communication. It is the foundation of everything we do. Nevertheless, communication is one behavior that managers have trouble developing. It is ironic that while many management classes and books talk about empowering employees to become comfortable speaking up, statistics do not support that employees truly feel the effect of this training. In one study, 70% of 1,100 employees and first-line supervisors across a diverse set of organizations, and 85% of subordinates in a different study, were uncomfortable or would not speak up about important issues such as managerial incompetence, decision-making procedures, inefficiencies, or poor performance (Carley 2013, 11).

In reality, many departments inadvertently reinforce poor communication by telling their firefighters or EMTs on the very first day that they need to keep their mouths shut and their ears open. Alternatively, imagine that on their first day we told them instead that they have been chosen to be members of the department because they are smart and talented people who are the future leaders of the department and as such are expected to perform as leaders. How would subordinates in your department respond about their comfort with speaking up? As indicated above, regardless of which approach they take, many department heads respond to this by saying, "Our people would be very comfortable speaking up," without actually asking their people to find out if that is true. We have found the above statistics to be consistently reflected within the fire service (Carley 2013).

Unfortunately, the fire service generally does not do well when it comes to subordinates speaking up to supervisors, regardless of whether it is in an emergency setting, because of the culture in which we develop (Kerwood 2008). This is because emergency services rely on a hierarchy to accomplish the required tasks. The hierarchy exists because emergency services are designed to respond to emergencies, but this carries over into the other 95% of the work done by emergency services. Rank, uniform insignia, hazing, and emphasis on seniority all contribute to reduced communication. It has been shown repeatedly that airplanes crashed when a crew member knew a problem existed but was afraid to speak up because of the hierarchy in the cockpit (Gladwell 2008).

The first place to start with improving communication is by diminishing the hierarchy. The most productive leaders make everyone in the group an equal (Orasanu and Fischer 2008; Reader, Flin, and Cuthbertson 2008). Picture King Arthur and his fabled Round Table. One can still be a strong leader while deferring to the group. This is done with a range of tools—from simple steps like avoiding titles to complex interactions like pointing out when the leader makes an error. A leader can improve communication by:

- Reducing reliance on symbolism (e.g., wearing the uniform insignia less)
- Avoiding use of titles and rank in normal conversation
- Admitting errors and bringing them up for discussion so that everyone can learn from them and hold other group members to the same standard
- Not hazing or diminishing others
- Acknowledging that the leader does only 20% of the work at most and that others in the group make it perform well 80% of the time

### Creating positive conflict

A combination of effective group development, the development of a cohesive team, trust, and strong communication set the stage for positive conflict. Such constructive conflict is critical to group success and is the best way to avoid any aspect of groupthink—when decisions are made without critical evaluation among the group because of fear of conflict. The fire service is particularly conditioned toward groupthink because of its reliance on hierarchical structure, directive leadership, and tendency to hand down orders without discussion (Carley 2009).

*Silence is disagreement by another method.*

*—Dane Carley*

It would seem logical that moving people between groups periodically would counteract groupthink. However, the opposite is actually true. Groupthink occurs because people with differing opinions or viewpoints are afraid to voice those views. Leaving a group together—particularly one that has been properly developed with officer guidance—actually encourages people to speak their viewpoints. In the norming and performing phases of a group, constructive criticism is freely given and prevents groupthink from happening. When people do not feel comfortable speaking their mind, as in underdeveloped groups, groupthink stifles performance and prevents group success.

## PUTTING IT ALL TOGETHER

The learning culture is the foundation for improving ourselves, our departments, and our profession. However, learning cannot exist unless the department's environment allows it to exist. Leadership is the environment in which the learning culture flourishes. This chapter, while short and succinct compared to the volumes of leadership information available, has elaborated on topics as important to improving ourselves, our departments, and our profession as any others discussed thus far.

# REFERENCES

Butler, C. J. 2005. "The Relationship between Emotional Intelligence and Transformational Leadership Behavior in Construction Industry Leaders." *Dissertation Abstracts International: Section B: The Sciences and Engineering* 66 (9): 4965.

Carley, D. A. 2008. *Impact of a 48 Hour Leadership Development Program on the Fargo Fire Department's Organizational Environment*. Unpublished manuscript. Department of Criminal Justice, St. Cloud State University, St. Cloud, MN.

———. 2009. *Organizational and Group Influences on an Individual in a Fire Department Company*. Unpublished manuscript. Department of Criminal Justice, St. Cloud State University, St. Cloud, MN.

———. 2014. *The Fargo Fire Department's Resource Investment in Community Risk Reduction Programs*. Emmitsburg, MD: National Fire Academy.

———. 2013. *Communication Impediments and Support in the Fargo Fire Department's Organizational Learning Process*. Emmitsburg, MD: National Fire Academy.

Edwards, S. T. 2005. *Fire Service Personnel Management*. 2nd ed. Upper Saddle River, NJ: Pearson.

Gladwell, M. 2008. *Outliers: The Story of Success*. New York: Little, Brown and Co.

Hogan, R. 2007. *Personality and the Fate of Organizations*. Mahwah, NJ: Lawrence Erlbaum Associates.

Hughes, R. L., R. C. Ginnett, and G. J. Curphy. 2006. *Leadership: Enhancing the Lessons of Experience*. 5th ed. New York: McGraw-Hill/Irwin.

Kerwood, S. D. 2008. *Identifying Barriers That Inhibit Institutionalizing Crew Resource Management in Combating Firefighter Casualties*. Minneapolis: ProQuest UMI.

Lines, R. 2005. "The Structure and Function of Attitudes toward Organizational Change." *Human Resource Development Review* 4 (1): 8–32.

Maxwell, J. C. 2004. *The 21 Indispensable Qualities of a Leader: Becoming the Person Others Will Want to Follow*. Purdue Lecture Notes Online website. http://www.ansc.purdue.edu/courses/communicationskills/leaderqualities.pdf.

Mullen, B., T. Anthony, E. Salas, and J. E. Driskell. 1994. "Group Cohesiveness and Quality of Decision Making: An Integration of Tests of the Groupthink Hypothesis." *Small Group Research* 25 (2): 189–204.

Myatt, M. 2010. *What All Great Leaders Have in Common*. N2Growth.com. http://www.n2growth.com/blog/the-learning-ceo/.

Oakley, E., and D. Krug. 1994. *Enlightened Leadership: Getting to the Heart of Change*. New York: Fireside.

Orasanu, J., and U. Fischer. 2008. "Improving Healthcare Communication: Lessons from the Flightdeck." In *Improving Healthcare Team*

*Communications: Building on Lessons from Aviation and Aerospace*, edited by C. P. Nemeth. Hampshire, UK: Ashgate Publishing, 23–45.

Reader, T., R. Flin, and B. Cuthbertson. 2008. "Factors Affecting Team Communication in the Intensive Care Unit (ICU)." In Nemeth, *Improving Healthcare Team Communications*, 117–133.

Soehren, M. S. 2001. *Perceptions of Managers Regarding Team Management Skills Used in Business*. Retrieved from Wayne State University Digital Commons (AAI3037148).

Stein, S. 2002. *The EQ Factor: Does Emotional Intelligence Make You a Better CEO?* Toronto, ON: Multi-Health Systems.

Zenger, J., and J. Folkman. 2009. "Ten Fatal Flaws That Derail Leaders." *Harvard Business Review*, June. http://hbr.org/2009/06/ten-fatal-flaws-that-derail-leaders/ar/1.

# 4

# WATCHING THE GAME TAPES TO PREPARE FOR NEXT WEEK

## CHAPTER OBJECTIVES

- Enhance understanding of near-miss reporting systems
- Explain how the data collected in near-miss reporting systems are analyzed
- Describe root-cause analysis

In 1812 Napoleon was a successful military leader who had amassed control over almost all of Europe. Although Napoleon had made an alliance with Russia, he set out to teach them a lesson for breaking his embargo with Great Britain. Given his large army, power, and past military success, Napoleon looked at Russia and saw an easy target. When he moved his armies into Russia, the Russians retreated to delay a direct confrontation with such a large army. In Russia, Napoleon's forces had to travel vast distances over barren lands and encountered food shortages and bitterly cold weather. The conquest began with around 500,000 troops and ended with about 20,000. A century later Hitler attempted a similar invasion that ended with similar results. In hindsight, historians might ask, "How did they not see that coming?"

Would it not be great if we all had psychic powers or a crystal ball that could tell us when and where the next fire service mistake, near-miss incident, or accident was going to come from? Would it not be great if we knew trouble was right around the corner? Although we do not have a crystal ball, we can still make pretty accurate predictions about where future mistakes may occur by learning from past mistakes. By learning about what has caused trouble for others in the past, we can tell where the next mistake, injury, or accident of our own is likely to cause us trouble tomorrow.

Harvard scholar George Santayana said, "Those who cannot remember the past are condemned to repeat it." This is not a cold hard fact—it's really more of a reflection—but there may be some truth to it. There have been military, political, and financial examples of people repeating mistakes already made by others throughout history. None of the examples are identical, but there are strong similarities. A more precise version of Santayana's quote might be, "Those who do not learn from the past mistakes of others are likely to repeat those mistakes." At an emergency scene, not only can mistakes be very embarrassing, but, more important, they can be costly—in terms of injuries, lives, public respect, public support, and property damage. We currently have the ability, the data, the technology, and the opportunity to learn from our past. We have the opportunity to learn about what has and has not worked for others at emergency scenes and why these things did or did not work.

All fires are different, but there are many similarities in our operations at all fires. By learning about what has caused others trouble in the past, we can tell where the next mistake, injury, or accident of our own is likely to get us in trouble. The good news is that we all do this on an individual level just by showing up at work, training, and responding to emergency incidents. We may not think about it, but on every one of these occasions, each of us is building our own experience. After many years, incidents, and training sessions, a firefighter acquires a lot of useful information. Some of us learn the lessons quickly, whereas others take longer to learn from their mistakes. Either way, we eventually learn from the experience and should be much better prepared the next time we encounter a similar situation. This process of learning occurs through experiences and is very important for us if we desire to be the wily veteran who seems to know everything.

But these experiences are limited to only firsthand incidents and training and maybe, if you are fortunate, the experiences of a few members on your own department who tell war stories or mentor younger firefighters. Basically, each of us is exposed only to the experience of one person or a few other people at best. What if you could gain the experience of many others and learn from it without actually having to make the mistake yourself? Imagine how much more you could learn and how much faster you could learn it. If you combined your own experiences with the experiences of others on your crew, others in your department, others in our country, and others around the world, you could build a vast set of experiences from which to learn.

## REMEMBER THAT TIME WHEN . . .

Near-miss reporting serves to categorize and collect experiences, mishaps, and incidents for learning purposes. To do so, we need to report them in an easily accessible, consolidated location so we that we can track and trend the data. Then, just as a hoseline can be followed toward the nozzle and the fire, the data collected will lead us to root causes and solutions. Did you know the tool we are speaking of is already on the apparatus? Many of you know of it, and some of you report on it; nevertheless, not enough data has been collected to begin forming solid trends identifying root causes of our incidents. We need everyone to know where this tool is and how to use it to get enough data to have a positive effect on the number of incidents, accidents, and fatalities seen in the fire service.

The tool we are talking about is the National Fire Fighter Near-Miss Reporting System. It has the potential to increase safety in the fire service dramatically, based on a similar successful experience—namely, near-miss reporting's contribution to airline safety. The National Fire Fighter Near-Miss Reporting System appeared in 2005, and in 2009 the system received 1,000 reports in a single year; still, many firefighters, officers, and chiefs have never heard of it or used it.

Approximately 1.2 million firefighters serve in the United States. If even 10% of the nation's firefighters each reported one near-miss per year, the fire service could begin tracking trends on 120,000 reports per year instead of the average of less than a thousand per year. This would be great if only 10% of us made one mistake each year, but the reality is that everyone makes numerous mistakes each year, so we should be able to achieve well over 1.2 million reports each year.

What is a near-miss? A near-miss is an incident that has the potential to cause harm or loss and almost did except for luck or quick thinking. Many of the best near-misses become crew stories. Near-misses are those stories that start with, "Remember that time when . . ." as someone tells it for the tenth time at the kitchen table. However, there is also something to learn from a near-miss that actually becomes an incident or accident. It is still a near-miss in the sense that maybe it could have been worse than it actually was, and it is an opportunity to learn. This is especially true when a near-miss cost someone their life; the best way to honor someone who gave their life should be to learn from it and make sure it does not happen again.

> *Many errors remain latent, embedded within the operational system until just the right combination of adverse events occurs.*
>
> —*Burke, Wilson, and Salas* (2005)

## THE NATIONAL FIRE FIGHTER NEAR-MISS REPORTING SYSTEM

Fortunately for us in the fire service, some of the most important organizations in our industry have placed a lot of time, effort, and money into making sure we have this valuable tool. In particular, when both the International Association of Fire Chiefs and the International Association of Fire Fighters say something is important, it is probably prudent to pay attention. The National Fire Fighter Near-Miss Reporting System website is dedicated to anonymously recording, tracking, and trending our fire service near-miss statistics through incident reporting. The reporting method removes all identifying information including names, cities, and apparatus numbers so that the focus is on learning from the incident instead of blaming individuals or departments. The reporting system is a place where we as a fire service can tell our remember-when stories around an electronic kitchen table for everyone to learn. The purpose is to improve safety in our industry by sharing information. The site's purpose is not for reports to be used by chiefs, law enforcement, or regulatory agencies (e.g., the Occupational Health and Safety Administration) to blame or punish individuals for their actions. The site also contains many other valuable resources such as training resources, searchable reports, and reports of the week that can be sent directly to your e-mail. Most recently, the near-miss program has been updated to provide a local near-miss reporting system that is linked to the national system. When a member submits a near-miss report locally, the department's designated reviewer can choose to publish it to the national system or not. Additionally, the system is being developed into a one-stop shop for submitting accident and injury reports and the associated paperwork specific to each employer and state worker's compensation system.

We have a great tool that is being used, but only at a portion of its capacity. Would you be fine with your engine's pump working at a portion of its capacity or your ladder stretching only 30% of the way? Why not? Is it because a poorly performing pump affects your safety and a malfunctioning ladder affects your ability to reach citizens or the roof? An inability to identify trends—and by extension root causes—similarly contributes to firefighter near-misses and accidents—an inability to perform at capacity. Several reasons contribute to why the site is not used enough to develop meaningful trends. They include:

- Fearing retribution or punishment
- Fearing harassment or teasing
- Not knowing it exists
- Not being familiar with how it works
- Not knowing what a near-miss is or when to report one
- Not wanting to do another report or generate more work
- Not understanding the benefits or impact of reporting

This is a fire service tool that, like a dull axe, has useful potential but is not being used as well as it should in training. We all know that tools do not work as well as they can and we are not as effective with them when we neglect to maintain and train with them.

Other industries use near-miss reporting as a critical component to learning, which has led to accident reduction rates as high as 80%. Therefore, fire service near-miss reporting, in conjunction with other components in a comprehensive fire service learning culture, can reduce firefighter accidents by up to 80% per year, based on the experience of other high-consequence industries using near-miss reporting, such as the medical field and airline industry.

Near-miss reporting, tracking incidents, and analyzing trends to identify a problem are fundamental to learning because problems persist when they are not recognized as such. The principles of a department striving to improve should include identifying potential problems early and addressing them. Therefore, participation in the National Fire Fighter Near-Miss Reporting System is an essential component of developing a learning culture. However, it is only one important component of a comprehensive plan for a department constantly striving to improve. There seems to be a relatively common desire to remain isolated from other departments, regions, and industries. Such an "our way is the only way" mentality increases hurdles to learning because it lowers the volume of data introduced to a person or agency. This means, at best, longer times to establish trends or, at worst, no trending because there is no information being shared from which to learn. Near-miss reporting, tracking, and trending attempt to identify human factors that are common denominators across all of the microcosms and regions of the country, thereby maximizing efforts to improve safety at a behavioral and attitudinal level instead of creating more rules to address the symptoms.

## BUT PROBLEMS IN MY REGION ARE NOT THE SAME

You may still be thinking, "A near-miss problem on the East Coast is not the same as a near-miss problem on the West Coast." When we look at symptoms, that line of reasoning may be true; for example, there are icy roads in the northeast but rarely in the southwest. However, we are talking about fundamental behaviors—the human factors we discussed in chapter 1.

For example, law enforcement uses considerably more psychological testing than the fire service and through this has found that over 70% of all police officers share similar personality traits (Hennessy 2007). The psychological research available in the fire service, though less substantial, tends to reveal similar trends. Margaret Ann Jensen (2005) has found supporting evidence that the majority of firefighters—68% in her study—are boredom-susceptible sensation seekers. Therefore, if we approach improving our safety on the job by addressing human factors, a nationwide trend is applicable, regardless of where the firefighter is, because the behavioral or attitudinal solution to a trend should address these 68% of the nation's firefighters assuming that additional personality research would reveal similar numbers. However, the abundance of similar evidence from other industries and their success rate in reducing incidents with this approach makes us confident that the fire service could see similar success.

## Exercises and Discussion Questions 4–1

How do you and your crew feel we can increase the use of the National Fire Fighter Near-Miss Reporting System? Do you believe fear of retribution is a factor preventing widespread use? Are there other factors?

Become familiar with the National Fire Fighter Near-Miss Reporting website. Go to the website and complete the following exercises:

- Enter a near-miss report
- Search the near-miss incidents
- Sign up for the report of the week by joining the mailing list

# ANALYZING REPORTING TO DETERMINE CONTRIBUTING FACTORS

Knowing the outcome of the decision after the incident is easy, so it is disingenuous and counterproductive to learning to say the decision maker should have foreseen the incident or its outcome. Root-cause analysis often finds that it is rare for a single person to make a bad decision intentionally or strive for a poor outcome. Almost always, human, organizational, and environmental factors each contribute a part to the poor decisions. In fact, we have found when teaching root-cause analysis of case studies that our classroom analyses frequently point to organizational influences, which are rooted in the fire service culture developed decades ago, and this finding is also supported by Kerwood's (2008) literature review.

Root-cause analysis stops the typical blame game to look for the true root cause, as the basis to learn from an incident. This requires acknowledging that we are all human and make mistakes. Even though we attribute 80% of our accidents to human factors, we usually focus on the 20% that are only the symptoms of a larger problem. We are not mathematicians, but would looking at the root causes encompassing 80% of our accidents not be more effective in reducing line-of-duty deaths and injuries? Why does the fire service ignore such a large category of causes of our accidents?

We know one reason is that the department or organization in which you work shapes how you see your world. Your reaction to an incident depends on your life experience, training, education, personal beliefs, and organizational experiences. A department striving to change should constantly challenge itself by asking critical thinking questions such as, "Are we seeing this accurately?" "Have we let our beliefs provide the answer to our questions without any evidence?" or, "Would someone from outside the fire service in another organization or profession see what happened the same way we do?" Critical, objective evaluation acknowledges the complexity of our problems but can be met with resistance because it denies the organization, the public, or the media the simple satisfaction of blaming a person, a piece of equipment, or lack of engineering when the true root cause probably lies within the organization's culture, leadership, training, disciplinary processes, firefighter personality traits, and public expectations. For example, the typical fire service reaction to a firefighter not wearing a seatbelt is to install another bell and whistle on the apparatus (treating the symptom). A more effective method is to shape the firefighter's behavior through identifying the root cause, setting an expectation, and teaching the importance of wearing a seatbelt (providing a solution)—and then holding the firefighter accountable for his or her actions in the future.

> *Don't blame the person for the error when the root cause of the error lies in the department's training, culture, and expectations.*
>
> —*Craig Nelson*

## FINDING A SOLUTION

To find true solutions, we first need to understand that a department's culture develops over years in response to administrative decisions, actions (or lack of actions), and training, which evolves into norms and values. Because the department is responsible for shaping the norms and values within a culture, the department heavily influences the employees' behavior. Departments striving to learn use this understanding to reduce the possibility of future accidents by performing root-cause analysis on accidents that do happen. After identifying root causes, the departments continue to track the reported data, build trend reports, and identify emerging trends as a holistic approach to safe operations because these departments understand that learning does not have an end.

Because humans operate in the department's system, accidents are inevitable. We all make mistakes. A near-miss error is a symptom of a larger systemic problem worth identifying and changing before it disables the organization. This constant desire to learn reduces the tendency to drift into complacency. Accidents are the result of the compounding of small errors, like links in a chain, leading to failure, often at an inopportune time. Performing root-cause analysis continuously as part of a learning culture helps firefighters and departments improve situational awareness by increasing their understanding of the organizational influences affecting their decision making (Carley 2010).

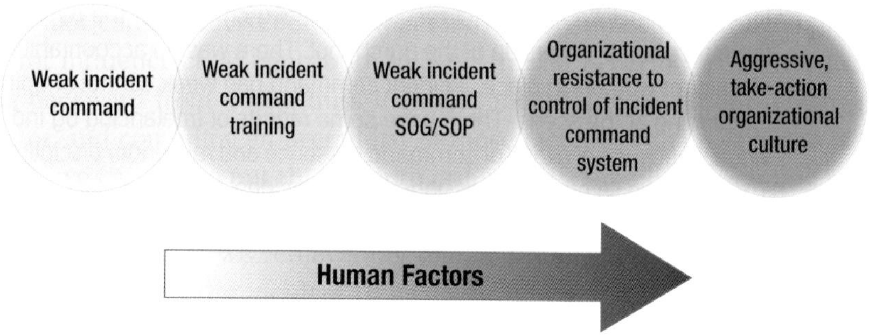


**Fig. 4–2.** A chain of errors working backwards from the traditional chain of errors to find the root cause of the problem, which almost always involves human factors

In the example above, there are other links preceding where we started the chain. For example, there would also be nonhuman-factors links that should be included, such as poor building construction, use of vinyl siding, etc. We chose to focus on the behavior (performing unsafe activities) leading to the injury because of the context of this specific discussion. The behavior leading to the injury was generally applicable across many different types of responses. The firefighters performing the unsafe behaviors are likely do so in other tasks, based on an aggressive, take-action organizational culture. It is also evident that the specific symptoms of unsafe behavior and weak incident command are part of a broader deficiency in the organizational culture influencing the behavior. The best part of performing a chain of errors in this manner is that it reveals the general behaviors creating an environment conducive to the unsafe behaviors. As a training officer, it becomes more effective to design a program addressing the fundamental behavior of being overly aggressive and the take-action culture because it is likely that personnel exhibit these behaviors on every type of call to which they respond (e.g., fires, car accidents, lines down, or gas leaks). By contrast, a program designed specifically for the incident described above, based on the lessons learned, would be more specific to a structure fire, particularly the building construction lessons. While building construction lessons will reduce the chance of a similar incident in the future, the lessons only apply to similar incident types.

## Real-Life Example

### Event description

We had a warehouse fire called in for smoke showing. When we arrived, my sergeant told me to pack out and see if we could knock it down. There was flame showing in the back right corner of the building and propane and welding equipment exploding and causing pop off valves to go off.

We decided to make entry to try to contain the fire and cool off the structure. When we tried to make entry at the side door, it flashed on us. We tried to fight it and back up. While in the doorway, about three feet back, we were still trying to fight it. I then noticed my air was hot and my face piece was starting to melt. We decided it was too late, and to start on a defensive attack.

When we backed out, we started peeling out of our gear due to being hot. The sergeant told me to stay there and he would be right back. He went to back the truck up due to radiant heat, and I only had on bunkers. Not paying much attention and due to miscommunication, he didn't return as he said. He thought I had returned to the front.

I then heard a loud explosion, and turned to look at the structure. Being only about 15 feet from the building, I saw a cloud of puffy smoke shoot out at me. I pulled my turnout jacket up, and the smoke and fire rolled over me. It then sucked back into the building, and I jumped up and ran. As I was running, I could hear the building falling. I fell about 10–15 feet away after running, and the wall hit right behind me. Needless to say, it scared me half to death.

I was picked up by a couple of rookies that were trying to get to me and saw it happen. They checked me out and got me away from it as I started to realize my arm had been burned through my jacket when I hid under it. My face had also been burned from the mask when it melted. I was treated and released from the hospital.

### Lessons learned

- This could have been prevented.
- We should have been together in teams of two to move the truck.

- I needed to wear my proper PPE while that close to the structure.
- We may not have needed to make entry with no lives at stake.
- In that situation, we all learned something from this event.

---

### Discussion Questions 4–2

In your experience reviewing accident investigations, do you, your department, and the fire service focus on the 80% cause of accidents (e.g., decision making or communication) or on the 20% cause of accidents (e.g., not following SOGs, lack of incident command structure at the emergency scene, equipment failure, or ignoring installed warning devices)? Where should we be focused? Why?

---

## USING ROOT-CAUSE ANALYSIS IN YOUR DEPARTMENT

Whether it is done locally or based on the National Fire Fighter Near-Miss Reporting System, reporting of near-misses and determining the root cause of the near-misses or accidents is only part of the solution. This data is useful, but how does a department put it to use? There are additional steps successful departments take to translate the data into value for those operating on the street.

After performing a root-cause analysis, the factors can be categorized. Categorizing the factors develops numbers that evolve into trends. A department may see equipment-related factors, for example, only once or twice a year in most years, but then equipment factors may take off the next year. Seeing this trend develop and having a time line of when the trend began developing helps isolate the cause. Sometimes, however, there may be a factor trending higher than the rest year after year. Communication is an example of a factor that appears repeatedly.

Even though this is a common factor, identifying the trends in its occurrence is important to see if this changes when a department tries different solutions.

There are three categories of factors to consider: common factors, severe factors, and emerging factors. Common factors may or may not be high risk, but they happen often. An increased exposure to the factor means an increased number of opportunities for an accident even if it is not a high-risk factor. Severe factors may or may not occur often, but they are high risk when they do with the potential for serious repercussions. Emerging factors are those that begin to trend up with a change in how the department operates. A change in technology, operating practices, or equipment causes unforeseen events to occur.

Once trends emerge, training—instead of more SOGs—should be developed to reduce the occurrence or severity of the identified factor. The goal of this training is to instill best practices into the department's operations by changing behaviors through an increased understanding. The best type of training for doing this is scenario-based training.

Effective scenario-based training incorporates realistic situations, is time compressed to increase the stress level without being excessive, and is completed in a learning environment where mistakes are expected but corrected appropriately. The goal is to incorporate changes into training prior to an accident, based on what is found from trending near-miss incidents.

## DEDICATED RISK-REDUCTION PERSONNEL

Completing all of the tasks described in this chapter may seem daunting. However, breaking the tasks down reduces the workload. The reporting falls on individual department members. The simpler and easier it is to access the reporting system, the more likely the reports will be completed. It is important to give an opportunity for members to expand on their response but to limit the required information to the most critical pieces. For example, a critical piece would be asking the member to report the factors contributing to the near-miss. The department is still able to analyze data for trends by using that little bit of information.

However, data collection, analyzing trends, and developing training also take time. The results of performing data analysis and trending improve as familiarity with what is being sought and the software used to analyze the data and trends increases. It is not uncommon to see resulting statistics begin with basic demographics and averages then evolve into targeted statistics that use quartiles and a combination of relevant demographics. It is helpful to have dedicated personnel for this reason.

Many departments have a safety committee that acts as the collection point. Having many minds analyzing the near-miss reports helps when doing a root-cause analysis because different people will see the same incident differently and recognize different contributing factors. Therefore, a safety committee should analyze near-miss reports as a group.

After root-cause analysis, task forces on lessons learned and error prevention, which could be subcommittees of the safety committee, can build training material from the analysis. The first task force reviews the root-cause contributing factors to develop lessons learned. The second task force takes the lessons learned and develops training programs to improve the department's understanding of these factors and prevent future errors. The training would likely include background information showing the trend, what contributes to the factors, solutions for addressing the factors, and training scenarios to practice these solutions.

## PUTTING IT ALL TOGETHER

Effectively reducing near-misses, which lead to accidents, is a multistep process. The goal is to gain a commitment from personnel to change behavior through training as the department learns. Gaining a commitment is in stark contrast to simply gaining compliance to a new rule and following the rule for fear of disciplinary action. The steps include:

- Implementing near-miss reporting or using the National Fire Fighter Near-Miss Reporting System to collect data from incidents for root-cause analysis

- Performing root-cause analysis to identify factors contributing to those near-misses:
  - The most common factors found in root-cause analysis provide high-frequency trends.
  - The most severe factors, common or not, found in root-cause analysis provide high-risk trends.
  - Emerging trends also provide those areas most in need of training because they are the most likely to be high frequency or high risk in the future. They develop because of new technologies, equipment, or procedures in our work environment.
- Developing training to address the identified factors by providing scenarios for personnel to practice the behaviors that will reduce the likelihood of an accident

## REFERENCES


Burke, C. S., Wilson, K. A., & Salas, E. 2005. "The Use of a Team-Based Strategy for Organizational Transformation: Guidance for Moving Toward a High Reliability Organization." *Theoretical Issues in Ergonomics Science,* 6 (6): 509–530.

Carley, D. A. 2010. "Working Paper on the Development of a Fire Department into a High Reliability Organization." Fargo, ND: Fargo Fire Department.

Hennessy, S. M. 2007. Interview by D. A. Carley, for Public Safety Executive Leadership 610: Advance Leadership Skills, September 13. St. Cloud State University. St. Cloud, MN.

Jensen, M. A. 2005. *The Relationship of the Sensation Seeking Personality Motive to Burnout, Injury, and Job Satisfaction among Firefighters.* New Orleans: University of New Orleans.

Kerwood, S. D. 2008. *Identifying Barriers That Inhibit Institutionalizing Crew Resource Management in Combating Firefighter Casualties.* Minneapolis: ProQuest UMI.

# 5

# CHANGE—THE ONLY THING AS SURE AS DEATH AND TAXES

## CHAPTER OBJECTIVES

- Illustrate the importance of being adaptable
- Discuss employees' concerns and reactions to change
- Provide methods of improving the acceptance of change

## LOSING THE EDGE

Chevrolet introduced one of the first commercially available gasoline injection systems in an American-made car. What year was it? It was 1957 when this became an option on the Corvette. What decade saw fuel injection become common for American-made cars? In the mid-1980s fuel injection became common on American-made cars. Thus, almost three decades passed between the introduction of fuel injection technology by an American manufacturer and its widespread adoption. Meanwhile, foreign car manufacturers, such as Porsche and Mercedes, embraced fuel injection far earlier. Pause and consider that information. Chances are good that your reaction might be along the lines, "Of course, those two companies used fuel injection and the

American companies didn't. That's because they are performance manufacturers." Are Porsche or Mercedes really better car builders than American car builders? The answer is no; however, they gained a reputation as such by appearing to be on the forefront of technology and by more readily embracing innovation and using it in their products.

> **Case in Point**
>
> As mentioned in an earlier chapter, Thomas J. Watson, an early leader at IBM said there was a market for about five computers worldwide (Maxwell, 2004). Fortunately, successive leadership changed the company's trajectory, and eventually they adapted to market demands. Still, the change did not come easy or fast. Rich Seidner (1996) once said that it took IBM at least nine months to ship an empty box because the company and its hierarchical structure adapted so slowly.

## BRINGING CHANGE TO AN ORGANIZATION

These stories of Chevrolet and IBM are good examples of how we change. Those in emergency services often joke about its resistance to change, but change does not come easily to any profession or organization. Regardless of our reputation, the emergency services are changing. These changes include:

- Adopting hazardous materials response
- Adopting technical rescue response
- Responding to EMS calls
- Positively adopting (not fighting) SCBA use
- Increasing the focus on fire prevention and community risk-reduction services
- Expanding participation in public education
- Increasing research and training in fire behavior and strategies and tactics
- Adding community paramedicine programs

## Exercise 5–1

First, make a list of changes that have happened in recent years in emergency services. Then, make a list of changes that you expect in the near future. Consider whether your department is part of these changes.

---

People naturally weigh change against how it affects them personally. This includes:

- What they expect to lose in the change
- Their comfort with the status quo as compared to the uncertainty of change
- How change conflicts with their values
- Resistance to a potentially increased workload

People often pass judgment quickly, before fully understanding the change. The natural tendency is to remain the same because it is comfortable (Heifetz and Linsky 2002; Bridges 2003; Lines 2005). Unfortunately, as Ed Oakley and Doug Krug (1994) have explained, an organization embracing the status quo is moving backward when compared to other organizations in the same industry that are progressing. Ridley (2010) has even provided evidence of people in history moving backward. Those isolated from others, by choice or by geography, not only stay the same but also begin to forget knowledge, skills, abilities, and technologies once known to them while retaining only their most frequently used skills.

The status quo is important to us because it represents a shared experience, our values, and our culture. Changing the status quo means not only trading away something that is comfortable but challenging the validity of the beliefs and values we have maintained throughout our lives. Therefore, challenging the status quo means overcoming immense, long-established hurdles.

It is easy to see that American car manufacturers continued producing products with which they were comfortable, maintaining the status quo of using carburetors rather than venturing into the unknown future of fuel injection. The push for American manufacturers to adopt fuel injection came from outside forces—namely government, through regulations, and foreign manufacturers, through increased competition from imported cars using fuel injection technology. Meanwhile, Porsche and Mercedes moved forward during the 1960s and '70s, improving on the fuel injection systems of the Bosch Company (another name associated with performance and innovation). Thus, American manufacturers, by maintaining the status quo could be considered to have gone backward in comparison to the European manufacturers embracing still-newer technology. In fact, Chevrolet even sold their fuel injection technology to the Bosch Company. This is similar to Ridley's (2010) finding that isolated people begin to forget what they once knew. Consequently, in the 1980s, Chevrolet needed to relearn what they had known about gasoline injection in the 1950s.

Therefore, it is not that the other manufacturers are better companies than American car manufacturers. Rather, they chose to accept the temporary discomfort of adopting new technology and challenging their beliefs and values because they envisioned a prominent reputation as a leader and a profitable future based on that reputation. Those leading the organizations overcame their personal hesitation to adopt the technology. They then led the employees through the transition period by communicating a vision with enough mastery that their employees overcame a sense of what was to be lost and instead focused on what was possible through innovation.

Bringing new ideas—change—to emergency services poses a similar challenge. Because this involves behavioral changes, we know it is not easy. Those who see a benefit from a proposed change will embrace it, while others will wait to see if it is merely another fad or if the idea has merit. Some will even form an opinion against an idea without first exploring it simply because change threatens the status quo. This is because change in emergency services challenges deep values and traditions passed from one generation to the next through our role models, teachers, officers, and other respected service members. Change only for the sake of change is detrimental to a department and its members, but the status quo does not move a department striving for success toward its goal or bring value to the community. Both of these models are precursors to a department's inevitable demise.

### Case in Point

The National Institute of Standards and Technology (NIST) and the Underwriters Laboratories Fire Service Research Institute have undertaken extensive research on fire behavior in recent years, and their results pose challenges to the fire service's traditional way of fighting fire. Even though this scientific research provides excellent data, some in the fire service have a difficult time accepting it because it challenges years of training that have evolved into personal beliefs of what is right. Because of this, some still try to make arguments against the research and its effect on our strategies that are based entirely on tradition. Ironically, the "tradition" of interior firefighting has existed only since the 1950s, when equipment provided the first opportunity to make an interior attack.

## MANAGING CHANGE IS A BALANCING ACT

Imagine that a pencil held vertically represents a department's progress forward: improvement through learning. The hand holding the pencil vertically represents the department's leadership guiding the process (fig. 5–1). The department needs to progress, because a static department is actually regressing when compared to other similar departments that are moving forward.

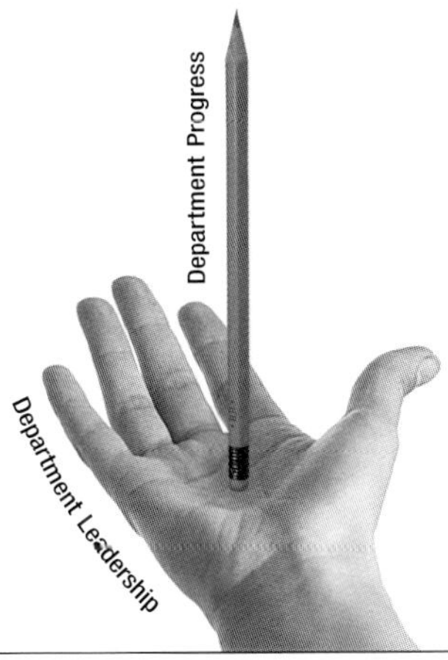


**Fig. 5–1.** A representation of leadership's hand balancing the pencil that represents the department's change

On the tip of the pencil is a square piece of paper. The square paper on a horizontal plane represents the department's balance. On opposite corners of the square are two department considerations. One corner represents the department changing too quickly, and the opposite corner represents the department changing to slowly. On the two remaining corners are considerations for the members of the department. One corner represents forcing change on members too quickly and the opposing corner represents waiting for members to change (fig. 5–2).

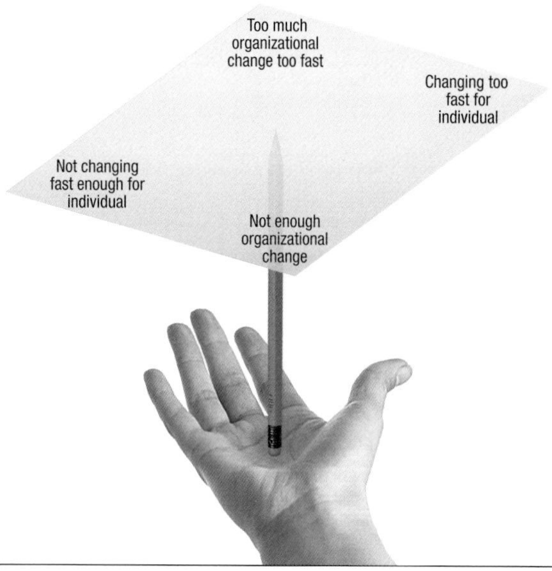


**Fig. 5–2.** A representation of the department's balancing plane on the tip of the pencil

Too much weight given to any one of the four corners or a combination of any two corners can cause the organization's balance to falter, and the square piece of paper can fall off the tip of the pencil, bringing the department's change—and morale—to a halt.

Taking this a step further, let us place 100 members on the square piece of paper. There is a natural arrangement of the 100 people because of human behavior and how it has been shown to emerge in most, if not all, organizations. About 10 of the members will be near the center of the square, right over the pencil's tip accepting and perpetuating the department's change. Another 10 of the members will be spread around the outside edge, some barely holding onto the department's balance by their fingertips. About 80 of the members will be spread somewhere between the center of the square, but outside of the 10 right in the middle, and inside of the 10 barely holding onto the edge (fig. 5–3).

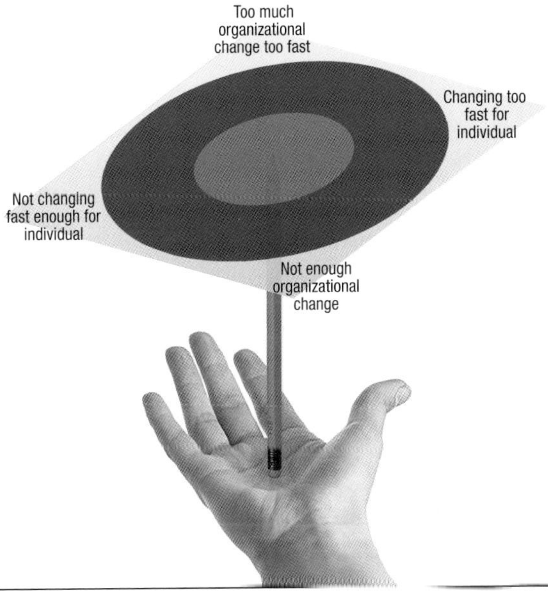


**Fig. 5–3.** A representation of balancing change. The dark circle on the horizontal plane represents the 80 members in the middle. The 10 perpetuating change are in the center circle, and the 10 slowing the department's change are along the edge of the plane outside of the black circle.

Ideally, a leader continues to raise the pencil vertically, sustaining change in the department, while balancing the square piece of paper, the department's balance. Not only does this help the department improve, but it also keeps morale positive because there is the right amount of change, for the right reasons, with guidance toward a clear goal. However, this if often unrealistic because one or two of the members barely hanging on the edge will finally let go as the rate of change or the type of change becomes unbearable to them.

In some ways this is unfortunate because the department lost a member who had something valuable—training, experience, or some type of other contribution to make. However, positive change is necessary for a department to continue being successful or to improve from its lessons learned. If a member cannot be led forward, then sometimes it becomes necessary, to maintain the department's balance, to let that person go so they can find a position in a different organization that better fits their expectations.

A leader who does not recognize that change is being forced upon one person too quickly can cause that person to move to the outer edge. This is particularly likely with the 80 members in the middle ring. Likewise, a leader may not bring change fast enough for another person, which causes them to move toward an outer edge. This is particularly likely with the 10 members in the inner ring but can happen to any of the groups, particularly if the middle group of 80 is disgruntled and wants to see change. Similarly, a leader can push a department too quickly or too slowly, or more importantly, without properly communicating a good reason or a clear goal.

One thing to recognize is that the balance can be thrown off by a combination of factors. Remember that the leader is balancing a square piece of paper on a point. If there are a few individuals who are beginning to become alienated and move toward the outer edge while the department moves that way as well, it can become impossible to balance the department as a whole.

The reaction to change differs among the groups. The group on the outer edge is going to complain about change almost every time. The group in the middle—the 80%—will possibly complain to some degree often but then get over it and move on. The 10% that are high performers will likely verbally support the changes even if they do not agree with them. However, even this group is reactive to the changes based on a combination of factors, which are discussed in the next few pages. It is typical for these employees to react the strongest when they do finally react and to demonstrate the most subversive behaviors when they have reached their tipping point.

## WHY IS THERE SO MUCH RESISTANCE TO CHANGE?

Being open to new ideas is a hallmark of leaders (Judge, Bono, Ilies, and Gerhardt 2002). Why then does there seem to be so much resistance to new ideas, which lead to change? Leaders are no different than other people. However, good leaders have learned to withhold their judgment on an idea until it is better understood. Humans, regardless of their position in life, tend to judge new ideas against one thing first: how the new idea will affect them personally. New ideas and change challenge the status quo. But what is the status quo? What about the status quo is being challenged by a new idea and change? New ideas and change challenge:

- Long-held beliefs and values
- What has been perceived as the best or right way in the past
- The person's involvement in perpetuating the current way of doing something
- A person's identity
- A person's future success based on past practices
- The ease with which a person conducts business as the new idea or change brings a different way of doing business that means new learning or increased work

Although ongoing change is the reality in most departments, changing a department's culture remains a challenge. Up to 75% of organizations' reorganizing efforts end in failure, the most frequently cited reason for which being neglect of the organization's culture (Stinchcomb and Ordaz 2007, 144). Partly this is because employees develop their work identity into their personal identity. "I am a firefighter," "I am a police officer," or "I am a paramedic" is a statement conveying the speaker's intense pride. Attached to that are the symbols of the position: the Maltese cross, the badge, the uniform, and the color of an apparatus or car. Psychologists consider our fear of change to be rooted in the loss of personal identity (Stinchcomb and Ordaz 2007).

A person's attitude toward a change also comes from how they feel the change matches their values and beliefs. Personal identity, values, and beliefs represent a person's core constructs, so considering how change affects people individually is important to the success of the change itself. Therefore, the process of implementing the change requires as much or more time in planning for the change and how it will be presented than for the change itself, because forming a positive attitude toward change at the outset is much easier than attempting to change a person's attitude toward change after it has been formed (Lines 2005, 21).

Developing a positive attitude toward change requires the change agent to consider how those affected by the change perceive it. A department's culture largely governs the relative status quo of an institutionalized organization. The culture comprises the employees' beliefs and values in action. From this, a psychological contract emerges, whereby employees come to have certain expectations of the employer's responsibilities and an understanding of what personal behaviors are most advantageous to their success. Discussion of change threatens the status quo—and, therefore, the psychological contract representing employees' expectations of what is fair and important.

Rune Lines (2005) has developed a framework of the many known variables that affect employees' attitudes toward change. Lines's discussion begins with two specific variables, job satisfaction and procedural justice, that a department controls to some extent in the design of the change process. Other variables that often play a part in acceptance or resistance are discussed subsequently.

## Job satisfaction

Change is weighed against values—particularly work values. These have been described by Nord and colleagues as "desired states that a person thinks he or she should be able to realize through working, and they guide a person's beliefs concerning specific jobs and work activities" (as cited by Lines 2005, 14). A belief, in this context, is not about religion or politics but instead about how a person sees the work environment—the amount of work, the type of work, the methods of work. This description closely resembles the psychological contract: a formal description of a person's expectations. When someone begins working in a department, they develop expectations of what the employer should provide. For example, those in emergency services place a high value on continuing education and training. The employees expect the department to support them in their continuing education because it benefits the department's ability to provide service and improves employee safety. Over time, this evolves from an expectation to an unwritten contract in the person's mind. Even though it is not a written and signed contract between the department and the employees, it is often viewed as carrying as much weight in the person' mind. This is a psychological contract. There is a strong reaction to anything that contradicts the psychological contract because it becomes a work belief, similar to a work value, in the person's mind.

A change that agrees with employees' work values and will provide a benefit contributes to a more accepting environment for change because the change feels like a reward to the employees. A department striving to agree with employees' work values by making department values and philosophy a part of the change has a higher likelihood of success in implementing that change. However, this in itself does not develop acceptance of change, nor does it help with changes that must happen but do not align with the employees' work values. Using procedural justice, discussed below, helps gain acceptance for changes that do not align with the employees' work values.

Lines (2005) has listed skill variety, task identity, significance, autonomy, and feedback as five job characteristics influencing an employee's job satisfaction. Skill variety is the opportunity employees have to apply their various skills; task identity is the opportunity to see a whole job to completion; significance is the degree to which the job affects the lives of others; autonomy is the extent of an employee's discretion; and feedback is the well-defined opportunity for employees to know their effectiveness (26). Because job satisfaction is generally high in emergency services, resistance to change represents the fear that change will affect one or more of those job characteristics. It also often reflects on how the process is handled instead of the change itself.

## Procedural justice

Chan Kim and Mauborgne (1993), Folger (1977), Konovsky (2000), and Korsgaard et al. (1995) have asserted that "how the change process is structured and carried out will influence attitudes towards change" (as cited by Lines 2005). Employees look at procedural justice as a measure of fairness in the department's processes. Processes that employees consider to be fair produce a positive attitude. Therefore, the relationship between a department's processes and employee attitudes is affected by the employees' perception of fairness (Lines 2005, 15). Fairness is introduced into the change process through procedural justice. Fairness in a change process includes the employees' abilities to "influence the establishment of rules for decisions," to provide "authentic opportunities to voice opinions," and to develop "systems of recourse" (Lines 2005, 15).

The ideal situation for creating fairness is to involve representatives of each group within the department in the change process. First, present the basic goals of a change's implementation to a group of employees from various background in terms of rank, years of experience, representative groups, and informal and formal leadership roles. Then, spend time describing the desirable outcome to build a shared mental model of the change among this diverse group. Last, guide the group in a change process of their own that achieves the vision and minimizes the negative results of the change. This multistep process increases the likelihood of success by incorporating procedural justice concepts. The various participants evaluate the proposals through the lens of their role in the department, allowing them to act as representatives of many others in similar roles and provide feedback on how the change will affect them. Therefore, each represented group within the department sees how the change process reflects what is fair to someone in a given position and develops buy-in for that group.

Lines (2004) found that participation in the change process increased the internalization of the change, reduced resistance to the change, and increased goal achievement. Participation provides a sense of control, which is a variable found to increase goal achievement (Lines 2004).

However, it is not always possible or practical to incorporate every group, person, or idea; this is especially true of large organizations, because groups that are too large lose their ability to function. Consequently, some may feel left out, or some concerns may be overlooked. In these cases, it is best to try to gather input with clearly defined parameters and limited responses such as in a multiple-choice questionnaire, which allows a large group to participate and give feedback that can be quickly analyzed in a spreadsheet and effectively used. Once the data are collected and analyzed, a smaller group can move forward with specific plans based on the feedback.

It also is important to define the group's authority and scope at the outset. Doing so sets clear expectations and boundaries. Not doing so sets the group up for failure and increased resistance to the change if the leader cannot adopt the process in its entirety.

### Case in Point

A 400-person department is interested in updating their mission statement. It is impractical to read and analyze 400 written essay suggestions. However, providing a multiple-choice questionnaire with value statements to prioritize in numerical order can be easily analyzed. From the top-scoring value statements, a smaller group can develop a mission statement for the department. By contrast, it may be more practical for a smaller department of 20 members to take 20 written essay suggestions and meld them into one mission statement.

## Other important variables

Social influence also affects employees' perceptions of a change initiative. It is likely that most departments initiating change experience varying reactions between duty shifts and even between stations, districts, or beats among the same shift. Rice and Aydin have observed that "peers, subordinates, and superiors in the individual's immediate work environment are likely to shape his or her interpretation" (as cited by Lines 2005). A leader should observe reactions and social interactions. By identifying the person or shift yielding an influence on employee's perceptions of change, the leader can likely find the root cause of resistance to a change, which likely will be a result of the culture or work environment allowed to develop within a crew by a supervisor (e.g., company officer, lead medic, or sergeant). This influence can be positive or negative. A person with a positive influence is a key leader, formal or informal, who can help with the change; a person with a negative influence is one who should be most involved in changes early on so that they buy into the idea and have a part in the change process. Although it may be tempting simply to ignore the people resisting the change and force them to adapt, this often ends unsuccessfully because it creates even more resistance.

> *The discussions that carry on at the kitchen table or briefing room are very influential in how change is perceived in a department.*
>
> —*Dane Carley*

Another variable shaping employees' perception of change is how the outcome affects their daily operations. Research shows that employees interpret change by weighing it against a relatively narrow self-interest first, rather than in terms of how it will benefit the department (Lines 2005, 14). For example, a change in data tracking that increases data entry is likely to receive greater resistance from company officers than firefighters because company officers typically perform the data entry. However, this is a simple example that likely elicits a relatively minor response. Changes to a promotional process likely would elicit stronger responses for the same basic reason but from a larger group of people. Although the change in the promotional process is intended to increase the chance of the department's success, this is a secondary consideration to the employees directly affected by the change, who may now need to learn a new system, align their education and training with new expectations, and compete with a different group of people who are suddenly better prepared than they were under the old system. Conversely, another group of employees may readily accept the change because the new process positions them to be more competitive for promotion than they were under the old system.

Social influences also shape employees' perceptions of the psychological contract between themselves and the employer. This affects their intake of information. Employees develop selective exposure, commonly known as "selective hearing." This essentially means that employees' minds will subconsciously select the information that best supports their beliefs and disregard any contradictory information. Because people better remember information that supports their position and overlook contradictory information, it reinforces their initial perception of whether a change either strengthens or weakens their position. Selective exposure is when someone remembers the information that supports their position, so selective exposure causes employees to embrace the change in the case of a positive perception, which is important to the department's success in the change process. But in the case of a negative perception, the employee will tend to reject and ignore the change, with the possibility of undermining the change process. Employees who generate a negative perception will overlook the positive aspects of the change and may only slowly, if ever, change their position (Lines 2005, 20). Selective exposure is one reason why forming a positive reception early in the change process is critical.

### Case in Point

Gordon Graham (2008) provides a great example of selective exposure. A Ford owner is driving down the road. The driver takes no notice of two Fords broken down on the side of the road. However, a mile or two further down the driver notices a Chevrolet broken down. The Ford driver automatically thinks, "Oh, look *another* Chevy broken down." This reinforces the Ford owner's belief that he will never buy a Chevrolet because they are always broken down. Similarly, a Chevrolet owner does the same thing when passing two broken-down Chevys only to notice the one broken-down Ford. In reality, we know that Fords and Chevrolets break down at about an equal rate and sometimes in a way completely unrelated to the maker (e.g., a flat tire). However, these owners are basing their perception on their beliefs that are likely influenced by their friends, family, and opinion instead of actual research or data, which reflects bias due to selective exposure.

Emotion also contributes to the strength of the reaction to the change process. This is because emotion affects employees' ability to gather objective information. In other words, a person with a strong, negative emotional response also displays strong selective behavior, making it even harder for a department to sell the positive aspects of the change.

## HOW DOES A CHANGE AGENT REDUCE RESISTANCE TO CHANGE?

A change agent is the messenger who seeks to get individuals to buy in to the change process through communication, someone who can effectively communicate the change process. This is important because "a key challenge for managing change is to control the attitude formation processes in the organization so that positive attitudes toward change are formed early in the change process" (Lines 2005, 24). Therefore, planning the change process becomes as important as—if not more important than—the change itself. Communication is the medium for conveying change, achieving the desired outcome, and shaping attitudes. If that is the case, how should a change agent communicate a new idea or change?

Effective communication begins with framing the change process honestly and relevantly. A change agent considers the change variables discussed thus far in the context of their department's culture, values, and philosophy of operations. A change agent must remember that they have already passed through the four phases of change—denial, resistance, exploration, and commitment (Lines 2005, 18)—when framing the message so that the agent considers the impact of the change from the employees' perspective. It is important to develop a message relevant to employees' concerns and position in the four phases of change to address their core concerns. In addition, a change agent is likely more committed to the change than others are at that point, so it is easy to overestimate others' involvement with the change (Lines 2005). Bridges (2003, 60) has recommended for change agents to consider four facets during the framing process:

- Explaining the basic purpose behind the sought outcome to help employees understand the logic
- Painting a picture of how the outcome looks and feels to help employees imaginatively experience the change
- Laying out a step-by-step plan for phasing in the outcome to provide employees a clear idea of where they need to go
- Giving each person a part to play in both the plan and the outcome to provide employees a tangible way to contribute and participate

Effective change communication includes (Lines 2005; Bridges 2003):

- Realistic descriptions of what the change will look like
- Expectation and acceptance of resistance and overreaction
- Open acknowledgment of the losses employees perceive
- Respect for successful past practices
- Acknowledgment that the proposed change is not necessarily perfect
- Disclosure that the change comes from many sources

- Association with positive emotions—or at the least, avoidance of negative emotions generated by portraying a doomsday scenario if the change does not occur

The strongest negative reactions evolve from changes perceived to threaten employees' core values and the identity they developed during their growth within the department (Lines 2005; Stinchcomb and Ordaz 2007). Employees generally view the past as successful, as aligned with their values, and as the source of their identity; therefore, any communication that demonizes past practices will foster negative feelings toward the change. Portraying the change as the solution to all problems compounds this reaction and casts doubt on the credibility of the messenger. Therefore, an effective messenger recognizes the successful processes of the past and acknowledges the potential hardships of the future.

Effective messengers build additional credibility and develop favorable attitudes by presenting "realistic change previews" (Lines 2005, 24). Negative emotional reactions increase when a change agent attempts to create readiness for change by "evoking an image of hostile environments that threaten the organization's existence if not acted upon" (24). Unfortunately, leadership books tend to perpetuate the idea that creating a crisis is an effective way to initiate change. However, avoiding negative imagery at the least and associating a positive environment (humor, manager's self-irony, ceremonies, etc.) contribute "strongly to positive attitudes toward change" (Lines 2005).

Last, enlisting various messengers in the discussion who employees perceive as independent sources will increase employees' motivation to think about the message (Lines 2005, 24). This part of the process is developed at the beginning by involving people already in the change process. The people who participated in the change process have passed through the four stages of change, have an interest in seeing it succeed, and understand the desired outcome.

## PUTTING IT ALL TOGETHER

Effective communication about a department's transition includes the formal and informal leaders of the representative groups of employees. When the message comes from these groups, it helps address relevant concerns of the department members because the messenger can identify with their concerns. The messenger should acknowledge the successes the department has had in the past and should specify how a change will improve the probability of the department's success in the future.

Unfortunately, employees do not consider the department's success first. Because employees weigh personal relevance first, the messenger must show how they too will be successful in the future. However, portraying the change as entirely positive builds skepticism and fuels negative attitudes. Change agents must acknowledge realities such as additional training (perceived as additional workload), changing expectations, and any other relevant consequence. A message focused on cost increases resistance and negatively affects internalization of the change (Lines 2004, 209), so even when cost is a consideration, it should not be the focus of the message. Then, the message is disseminated widely, repeatedly, and by the many formal and informal organizational leaders who hold meaningful perspectives of the change.

## MANAGING THE TRANSITION

As during the introduction of a change, competently managing the transition improves the likelihood of success ultimately. Moreover, effective communication is key. Bridges (2003) has recommended using a transition management team to communicate about the change and manage the transition. Such a team proved to be invaluable in the successful integration of a sheriff's office and county fire department (Stinchcomb and Ordaz 2007).

A transition management team includes seven to twelve people who represent as wide a cross section of employees as possible while excluding members of management or members who previously participated in the change process. Members of management or members who were part of the change process can have hidden agendas because of a personal attachment to the change.

Administrators must clearly define the team's expectations. This does not include any decision-making authority because the team is intended only as a communication-enhancing entity. The main roles of the team are to monitor employees' concerns and relay them directly to the management staff, to review plans or communications to ensure employees perceive messages as management intended, to counter rumors, and to verify that management responds to concerns. It is important that this team work in an environment where they feel free to speak because it is likely that these roles will expose them to concerns that should be voiced to the administration but may challenge the administration's position.

The aforementioned transition management team operating in the merger of a sheriff's office and county fire department had a particularly challenging job. Few known successful law enforcement and fire department mergers exist. In this case, the organizational culture of the two agencies was a prime consideration. The main focus of the team was to align the established cultural values and identities with the new organizational structure. Although the specific concerns of a merger are not relevant to transitioning an emergency services department in every case, the importance of aligning employees' culture, values, and norms is important in any transition.

The transition management team acts as an information conduit between employees and administrators keeping people informed. In the law enforcement–fire department merger described by Stinchcomb and Ordaz (2007), among those interviewed, "good communication between transition members and employees was key. . . . There was no room for rumors" (152). Additionally, the department established a phone line to answer questions, which also acted to reduce rumors. This action positively influenced employee perceptions shaped within the smaller group settings of widely dispersed fire stations because the phone line squelched rumors, and an entity outside the immediate circle of change agents reinforced organizational communication.

The announcement of a change threatens employees' work identities, organizational culture, and accepted norms. The transition period is turbulent because employees begin moving away from the old norms and identities toward new norms and identities. Confusion exists because the transition period is undefined, compared to the status quo of the past. A properly managed transition:

- Has short-term goals to provide opportunity for success
- Maintains continued communication about the future

- Does not introduce additional change before employees finish the original transition period

The transition period ends as employees accept the new identity, organizational culture, and norms (Bridges 2003).

## REFERENCES

Bridges, W. 2003. *Managing Transitions: Making the Most of Change*. Cambridge, MA: Perseus.

Graham, G. 2008. Cognitive Dissonance Theory. http://youtu.be/HjU-vtPS1nw.

Heifetz, R. A., and M. Linsky. 2002. *Leadership on the Line: Staying Alive through the Dangers of Leading*. Boston: Harvard Business School Press.

Judge, T. A., J. E. Bono, R. Ilies, and M. W. Gerhardt. 2002. "Personality and Leadership: A Qualitative and Quantitative Review." *Journal of Applied Psychology* 87 (4): 765–780.

Lines, R. 2004. "Influence of Participation in Strategic Change: Resistance, Organizational Commitment and Change Goal Achievement." *Journal of Change Management* 4 (3): 193–215.

———. 2005. "The Structure and Function of Attitudes toward Organizational Change." *Human Resource Development Review* 4 (1): 8–32.

Maxwell, J. C. 2004. *The 21 Indispensable Qualities of a Leader: Becoming the Person Others Will Want to Follow*. Purdue Lecture Notes Online website. http://www.ansc.purdue.edu/courses/communicationskills/leaderqualities.pdf.

Oakley, E., and D. Krug. 1994. *Enlightened Leadership: Getting to the Heart of Change*. New York: Simon & Schuster.

Ridley, M. 2010. *The Rational Optimist: How Prosperity Evolves*. New York: Harper.

Seidner, R. 1996. Interviewed in *Triumph of the Nerds: The Rise of Accidental Empires*. By R. X. Cringely. Season 1, part 2. PBS.

Stinchcomb, J. B., and F. Ordaz. 2007. "The Integration of Two 'Brotherhoods' into One Organizational Culture: A Psycho-Social Perspective on Merging Police and Fire Services." *Public Organization Review* 7 (2): 143–161.

# 6

# HIGHER RELIABILITY ORGANIZING

## CHAPTER OBJECTIVES

- Connect learning and leadership to operating reliably
- Introduce five higher reliability principles to guide the organization
- Identify eight higher reliability employee behaviors to develop in the people working in the organization

## OPERATING RELIABLY

We hope that you are thinking, "Wow, this sounds great! How do we do all of this consistently?" Even though it may be easier than it sounds, it will still take a lot of work. The reason it is easier is because a process already exists for the fire service to follow, called higher reliability organizing. This process is based on two models of higher reliability organizations (HROs): the first is based on five organizational principles, and the other on eight employee behaviors. Nevertheless, the process is a lot of work because an HRO never accepts the status quo as adequate; an HRO is always striving to improve, striving for success.

We have described in depth the ways a department can strive for success. Have you noticed an underlying theme so far? Who—not what—is the focus of most of the content. A department striving for success focuses on the people in the department and how it can support them in their tasks. This is because HROs recognize that it is the people in the organization who make it successful and make it highly reliable. HROs have reviewed the data and trends and seen that machines do not fail very often; when a near-miss or accident occurs, it is the result of people and how the organization trained and supported those people 80% of the time (International Association of Fire Chiefs 2003).

We have spent a lot of time explaining what amounts to leadership and how people react because this represents a fundamental aspect of HROs. Instilling leadership in the culture and focusing on the people allows an organization to optimize the five principles and eight behaviors. In fact, an HRO cannot exist unless these core elements are integrated into the department. The five principles and eight behaviors are part of a system that helps the department function consistently. An HRO brings attention to and allows development around the human factors, learning culture, and leadership traits discussed thus far.

## INTRODUCTION TO HIGHER RELIABILITY ORGANIZATIONS

Fire service leaders have long sought answers to why the fire service loses about 100 firefighters annually in the line of duty. Two areas of focus have been a goal of reducing line-of-duty deaths (LODDs). The first is reducing LODDs by 50% between 2004 and 2014 and the second is implementing the 16 Life Safety Initiatives to guide the fire service toward that goal. Years later, the fire service still sees about 100 firefighter LODDs annually. Why does the fire service continue to lose roughly the same number of firefighters? It seems like every time we turn around someone is preaching safety, yet the numbers stay the same. Again, why is that? We have intelligent, inspiring leaders who recognize the problem and provide solutions. We have good firefighters and company officers who know their jobs inside and out. They follow the rules, use solid safety practices, and train more often than ever before. Leaders, firefighters, and company officers alike are well trained, well educated, and experienced. Therefore, we believe the problem lies within our approach to safety. We need a different approach to make significant headway.

The 16 Life Safety Initiatives advocated by *Everyone Goes Home* represent a nice departure from many of our safety programs because the initiatives explore a short list of root causes, in contrast to previous safety lists that are too long and complex to remember. Some lists contain symptomatic problems like "Wear your seatbelt" and "Put on your SCBA" instead of addressing the root problem or behavior. The problem is that 80% of the reported near-misses come from human behavior, not from the symptoms of the behavior. Therefore, we need a comprehensive system that addresses the fundamental behaviors contributing to LODDs and injuries. This means a change in our approach.

Other industries operating in an environment with severe consequences for a single mistake do not see similar accident statistics. What do they do differently? For one, these industries have aggressively implemented comprehensive systems focused on human solutions that have reduced fatal accidents and injuries by up to 80%. Imagine up to 80 fewer line-of-duty deaths and 80,000 fewer firefighter injuries. Is it possible? We believe it is!

Such a dramatic reduction will be possible if we shift our focus from merely treating the symptoms causing line-of-duty deaths and injuries and look at the root cause instead. In this chapter, we will lay the foundation for a comprehensive system that addresses operations, prevention, and quality control using components the emergency services already have in place. Industries already using this type of system refer to it as an HRO.

Some people and researchers already consider emergency services an HRO. The definition of an HRO is an organization operating in a complex, high risk environment in which a single error has the potential for disastrous consequences yet the organization routinely performs with a low number of errors due to various organizational characteristics intentionally engineered to prevent human error that will "continually delay or even permanently defer the inevitable failures" (Erickson and Dyer 2004, 8). An HRO accomplishes this by deliberately implementing a comprehensive plan that hinges on developing a learning culture within an organization using methods such as near-miss reporting and root-cause analysis.

## FIVE OPERATING PRINCIPLES FOR THE DEPARTMENT

An HRO recognizes that departments are composed of humans and that humans, no matter how diligent they may be, make mistakes. In the fire service, we must first accept that we all make mistakes, no matter how smart, educated, or talented we are. An HRO builds systems upon five basic principles to:

- Prevent mistakes
- Compensate when mistakes do occur
- Learn from any given mistake to prevent it from occurring again.

Weick and Sutcliffe (2007) have listed the principles of an HRO as:

- Preoccupation with failure
- Reluctance to simplify
- Sensitivity to operations
- Commitment to resilience
- Deference to expertise

We have translated these five principles into terms specific to the fire service:

- Seek constant improvement
- Avoid oversimplification
- Listen to those working in the streets
- Adapt successfully
- Listen to those with expertise

These principles are evident in fire service systems such as the incident command system (ICS), in particular FIRESCOPE ICS, and quality-assurance (QA) programs associated with advanced life support programs. However, these systems are in silos instead of being integrated components of a HRO system. An HRO incorporates multiple systems including ICS and QA programs along with crew resource management (CRM), understanding human behavior, and leadership into one comprehensive system. Therefore, all of these become gears working together in the HRO machine.

## Seek constant improvement

The political climate in the United States pushes the fire service to provide an effective service efficiently. Fire departments will not be able to exist in the future as they do today because the citizens' definition of a professional fire department capable of meeting their needs is constantly evolving. We believe that the principles of an HRO address not only the needs of firefighters by creating a safer work environment but also the citizens' desire for the fire department to continually evolve with their expectations.

Preoccupation with failure is the technical term given this principle by Weick (1990). Granted, a preoccupation with failure sounds terribly negative at first hearing. However, it acknowledges that a fire department has not experienced all of the ways that their systems can fail, so a department should be focused on possible future failures.

**How it applies.** In the fire service, we should seek constant improvement to avoid potential failures. We must recognize that we cannot know all the possible ways we may fail in the future because of the dynamic and ever-changing nature of emergency response (Weick and Sutcliffe 2007, 3). Simply put, we will all encounter more accidents in the future. Firefighters understand this concept but resist accepting it for obvious reasons.

Seeking constant improvement means we must question everything. By questioning everything, we can discover the potential failure points in our routines. An established routine is a solution only to the last problem the department experienced, so it will not necessarily solve a future problem. Although routine is necessary in skills (e.g., donning SCBA or pulling preconnects), when routine is ingrained in thought processes and decision making, it reduces our ability to:

- See potential failures in our practices
- Identify possible alternative ideas
- Evaluate other solutions

One method to reduce routine in our thought processes and decision making is to collect accurate data and analyze our operations objectively. As humans, we naturally seek out evidence supporting what we believe, regardless whether it is true (i.e., we follow our gut feelings). By analyzing our operations, we seek constant improvement, thereby reducing our failures. HROs practice this principle by:

- *Recognizing discrepancies* and exploring them instead of mindlessly categorizing them in familiar, comfortable places
- *Questioning an error* over and over to find the root cause, which often lies in a fire department's culture and practices
- *Collecting after-action reports quickly* before stories of the various participants are "straightened out" to match what they wished happened
- *Reducing distractions* through delegation to maintain situational awareness
- *Encouraging an unsimplification of operations* by seeking out different perspectives, people with different backgrounds, and constructive, respectful feedback

- *Doing away with rigid hierarchical decision making* and replacing it with value-based decision making that is dynamic and flexible

Weick and Sutcliffe (2007, 47) and Driessen, Outka-Perkins, and Anderson (2005) have also found many common behavioral contributors to failure. These traits are flags that signal a high potential for failure. The flags the fire service must watch for include:

- Recent changes in supervision
- Issues delegated without follow-up
- Lack of questioning attitude
- Missed steps in a procedure
- Loss of collective situational awareness; not everyone is on the same page
- Staff spread thin
- Distraction

**Traits.** There are three emergency services traits that can help or hinder our ability to seek improvement, depending on how they are used. These three traits are:

- Systems
- Emergency services hierarchy
- Task overload

*Systems.* So how does the fire service seek constant improvement to provide effective services? Fire departments already have systems in place to address many aspects of the first principle; however, critical components of those systems are often lost. Just because a system is being used does not mean it is being used as intended. This may happen because the users misinterpret how it should be used, selectively use only parts of the systems, or use it only on occasion. While we discuss these systems in greater depth throughout the book, a short list of some of the systems that help us seek constant improvement but can fall into one of the categories of misuse are:

- CRM
- Near-miss reporting (National Fire Fighter Near-Miss Reporting System)
- ICS
- Root-cause analysis
- National Fire Incident Reporting System
- Accountability
- Rapid-intervention crew

*Hierarchy.* We must also refine our hierarchical structure to seek constant improvement. Even though hierarchy is necessary to divide the workload and accomplish tasks, input from citizens and government leaders is critical on daily operations, and input from even the newest firefighter should be encouraged—rather than simply expected. Fire departments should also invite input and opposing views at an administrative level to ensure that the department is providing an effective service efficiently that meets citizens' expectations. Likewise, incident commanders and company officers should expect firefighters to speak up whenever they feel something may be wrong.

The fire service spends a great amount of time and money weeding out candidates to develop a list that represents the cream of the crop. We hire firefighters with education, training, and experience who have proven through testing that they are the best—then we tell them not to speak because they do not know anything. Regardless of our best intentions, this behavior transfers from the training ground and fire station to the emergency scene. It is possible—in fact it is likely—that new firefighters will bring valuable contributions and catch details even the best fire minds in the business may have missed. An HRO seeks out differing views and incorporates them into the fire department's operations to improve effectiveness.

*Task overload.* It is admirable and necessary for firefighters to maintain a can-do attitude. Nevertheless, this attitude can lead to problems because we sometimes try to succeed regardless of how short we are on resources or how dire the situation is. The fire service's can-do attitude and positive values sometimes cause firefighters to try to accomplish more work or take more risks than is realistic or appropriate to a given situation. The political climate that fire departments operate within nationwide pushes for more efficient staffing; however, efficient staffing that is no longer effective complicates this by reducing our ability to do our job safely and effectively. A driver who is trying to place the apparatus, put the pump in gear, pull lines, flow water to the right line at the right pressure, establish a water supply, and place a PPV fan does not have time to look around the scene and consequently may miss important cues. Similarly, incident commanders are trying to monitor multiple radio channels, track resources, and make strategic decisions simultaneously because an aide is not available to assist. This task overload leads to:

- Delegating without adequate follow-up
- Missed cues or flags
- Distraction because our minds cannot multitask
- Missed steps in a procedure (e.g., accountability)
- Inability to develop an incident action plan and share it with crews to improve collective situational awareness
- Reduction of feedback during the decision-making process

## Real-Life Example

### Event description

Just after midnight, a call was received from an occupant of a large building asking for assistance in securing a fire alarm, which was going off. I got my driver and medic crew and proceeded to the building without notifying the shift commander, as I should have. Upon arrival at the facility we were met by a worker stating that there had been a smell inside for a few days and now the alarm had activated. He reported seeing no signs of smoke or fire as he left the structure.

My driver and I proceeded into the building with the civilian as an escort, due to the need for keyed entry throughout the building. Upon entry there was an audible alarm as well as a visual notification. We proceeded up to the second floor and into the computer area where the smell had been reported. At this time a haze was noticed at the ceiling extending down approximately two feet. The civilian said that when he left the room it was clear. My partner and I looked around the room for possible causes of the haze when we began to feel heat coming from behind us. I turned to my partner to tell him we needed to vacate the building when I noticed my voice had deepened and slowed considerably. I knew at this time we had been exposed to something.

A full commercial response was struck upon evacuation of the structure. The building was searched and found to have a circuit breaker taped opened as well as having had a release of FM 200 in the computer room (an agent much like Halon). All crew members were medically cleared and felt no ill effects. The structure was released back to the occupants upon completion of operations.

## Lessons learned

I learned many lessons from this near-miss incident. The first of which was a lack of communications to the shift commander. His lack of notification of where crews were and in what manner they were operating was cause for surprise and delay. Further learning occurred from not following the given departmental policies on alarm responses and staffing for entry into a structure with an activated alarm. While there were four personnel on scene, policy further states that a rapid intervention team is to be established where there is no imminent lives in peril. I showed a complacent attitude in not being properly prepared to enter the structure, establish command, and communicate to those outside. Another lesson learned was not to be disarmed by the civilian understating the possible problem and allowing myself to be lulled into a false sense of security.

Lessons were also learned in regards to the building itself. Being a secured government building, we had no real plans to identify possible hazards within the structure as well as the ability to access areas throughout. The light above the computer room door, identifying there had been a release, was not activated.

## Discussion Questions 6–1

Which bulleted points preceding this Real-Life Example are relevant to the incident described? Are there other ways to seek constant improvement beyond those listed above?

---

### Avoid oversimplifying

In such a complex, rapidly changing world, a reluctance to simplify seems counterproductive. We live and work in a complicated world. We also know that complicated environments are rife with opportunity for mistakes because their inherent complexity makes them difficult to understand, leading to a lack of situational awareness. Therefore, simplifying reduces the chance for error. Or does it? An HRO simplifies but does so reluctantly. Why? This is because an HRO realizes that simplification means generalizing, which contributes to complacency and reduced situational awareness.

Simplifying tasks can mean increasing unfamiliarity. Weick and Sutcliffe (2007, 39) have observed that "it is impossible to manage any organization solely by means of mindless control systems that depend on rules, plans, routines, stable categories, and fixed criteria for correct performance." Thus, avoiding oversimplification means:

- Understanding that oversimplifying limits exploration, experience, use, practice, and ability
- Recognizing that creating general labels limits understanding
- Using checklists incorrectly; relying on them entirely without ascertaining how it fits the current situation
- Challenging the status quo with constructive conflict

For the fire service, a reluctance to simplify should mean to avoid oversimplifying. An example of oversimplifying can be drawn from our regular equipment checks. We check our medical bags and equipment daily, which leads to a common simplification of placing ties on zippers of compartments used less often. If the tie is not broken, then the compartment is ready so no one opens it. This definitely increases our daily productivity to make us more efficient, but the routine reduces the number of times firefighters put their hands on their tools, so they become less effective on scene. We begin losing our situational awareness as we become less familiar with what is in each compartment. This is a simple example, but it illustrates the basic premise well.

A fire department avoiding oversimplification finds a balance between daily efficiency and providing effective services by:

- Encouraging crews to put their hands on their equipment during daily checks
- Placing the appropriate weight on each person's ability, experience, and education to fulfill the roles they are assigned to perform (e.g., moving up to act as a driver or officer)
- Remembering to connect the reason for fire prevention codes to the inspection process
- Recognizing that there is more than one way to do something (e.g., moving beyond the idea that there is only one right way to put up a ladder, ventilate, fight a fire, use a hose stream, etc.)
- Encouraging diversity in personnel, opinions, and backgrounds
- Encouraging their members to ask why

> *A routine is only a solution to the last problem the department experienced but the next problem will likely not be solved by the same routine. Routine may actually inhibit finding a solution to a future problem by narrowing the ideas people develop.*
>
> —*Dane Carley*

**Labels.** It is important to recognize that creating labels limits understanding. Labels are ways for us to simplify conversation. However, labels are general terms that can be understood differently by different people. Think about the brand name Coca-Cola. Depending on where you live in the country, the same beverage could be called Coke, pop, or soda. Generalization, especially when used on emergency scenes, can lead to reduced situational awareness and confusion.

This problem is magnified at scenes with more than one agency or department. Examples of labels that can lead to problems depending on location or experience include:

- Tanker/tender
- Truck company/ladder company
- Engine/pumper
- Smoke showing (heavy or light?)
- The human mind/checklists

**Checklists.** The human mind is not adequate for complex lists of procedures or operations, so we develop checklists. The problem is that we may become overly reliant on checklists, and the meaning of each of the checklists' individual components gets lost. Using either the human mind or checklists separately is oversimplifying. The human mind alone has the understanding but is forgetful; the checklist alone remembers but does not have the understanding. A person reading a checklist needs the human mind to understand what the meaning of the components in the checklist represent.

## Case in Point

There are many cases where we look at an incident and recognize more than one point where a different decision would have caused different outcomes. When we start at point A, we may not know how many different outcomes are possible because of all the decision points between point A and the end point. The Cerro Grande fire is one such example. The National Park Service's (2001) board of inquiry issued a report examining causes contributing to the prescribed fire's growth into an uncontrolled fire that ultimately burned 235 homes around Los Alamos, NM.

Their report states that a combination of system flaws and personal decisions contributed to a chain of events that led to the wildfire. The report details 23 separate decision points, plus several decision points within those 23, that placed a link in the chain of events leading to the wildfire. The first three contributing factors center on the use of a rating system designed to determine the complexity of a prescribed fire. It is presumed that a simple website design error incorrectly numbered the complexity scale the burn boss used to measure the fire's complexity. Because the burn boss was not entirely familiar with the complexity rating process or the normal labels and their meaning, the burn was consistently underrated for its complexity according to the report. This led to some critical decisions: first, to ignite the prescribed burn; second, to release resources; and third, to develop a weak contingency plan.

Labels in a checklist simplify decision making and provide reminders for steps that should occur. However, it is critical that those using the checklist understand the meaning of labels in the checklist and their function. Such understanding provides situational awareness and contributes to sound decision making. Unfortunately, not understanding the meaning of the labels or their underlying function, as in this case, can contribute to significantly poorer decision making (National Park Service 2001). Airlines, on the other hand, learn their aircraft through intensive training and use checklists only as memory aids.

---

**Asking why.** Asking why is a straightforward way to avoid oversimplifying. This is because it develops constructive conflict, which is fundamental to simplifying for the right reasons. Asking, "Why do we do it this way?" and receiving the answer, "Because that is how we've always done it," is not sufficient and should raise a warning flag. Instead, the answer should be because the way we do it is safer, more efficient, or more effective. Even receiving the answer that it is more fun, faster, easier, or cheaper is acceptable if it does not affect safety or the quality of service. There only needs to be a good reason for why something is done a certain way. As an example, imagine this conversation:

> Son: "What is the best way to cook a roast, Mom?"
>
> Mother: "You start by cutting off both ends of the roast for the best results."
>
> Son: "Why do you cut both ends off of the roast?"
>
> Mother: "Because Grandma always did."
>
> Son [now to Grandma]: "Why do you always cut both ends off of the roast? Does it make it better?"

Grandma: "No, I just don't have a pan big enough to fit the roast without cutting the ends off."

This humorous scenario shows why it is so important to ask why in order to improve a fire department's operations. However, the grandmother does not operate in a life safety environment, so as ridiculous as the previous conversation sounds, it is even more critical for those of us in the fire service to use constructive but positive criticism in our constant effort to improve. A diverse employee pool also improves this situation because various backgrounds, education, training, opinions, and experience bring different why questions. In essence, it is avoiding the "moth to the candle" syndrome at a command and administrative level by being open to questions from the bottom up.

## Real-Life Example

### Event description

Units were dispatched to an apartment fire reported "in the area of" with no address. The battalion chief arrived on scene and communicated a working fire. I was the officer on the first arriving engine. We found four apartments with heavy fire involvement and command advised us to hit it from the other side. Not knowing where the other side was, we changed from pulling a 3" attack line to establishing a hose lay attacking the fire with a 1¾" line. Command advised we had a second crew coming in behind us. We attacked fire on the first floor, knocking down major portions of fire in the first two units. My crew advanced the line to the second floor for fire attack. During this time the fire began to intensify. The second crew was delayed in advancing the second line to the first-floor units.

While completing attack on the second floor, the floor collapsed, causing me to fall into the first floor. My two firefighters, who were exiting the building, advised command of the incident. Command continued communicating over the radio. I was unable to call a Mayday because of the radio traffic. I rescued myself out of the first floor and attempted to locate my crew. Command had advised them to go get me. One went inside and one went around the back. After not finding my crew, I found command and advised him I was out and trying to locate my crew. We exchanged words and I called a Mayday declaring a lost crew. There were no RIT [rapid intervention team] or back-up crews. I also advised command to go into "defensive mode" and call for a PAR [personnel accountability report].

After several tense moments, my crew was located. There was a failure of an on-scene report advising crews of location and conditions. Failure to identify, properly state task assignments, and a failure on my part to question command on my assignment to "attack from the other side."

The first crew was aggressive making it to the second floor; I did not check to ensure fire was in control prior to advancing above.

### Lessons learned

There was a failure to have a RIT or backup units in place to assist. A good command system should have been established from the beginning. Staging should have been established with the amount of fire we had and the building construction. There was no department review or critique of the incident. Command believed it was a lack of proper actions by the first officer. I accept my mistake and have taken action to improve my abilities. The department should have conducted an investigation and a post-incident analysis so everyone could learn from the incident.

---

## Discussion Questions 6–2

1. The fire service teaches its members from day one not to question orders at an emergency scene. This is a cultural issue across the entire fire service. How did this culture contribute to this event?
2. Develop a chain of errors for this event. Does your chain of errors lead to cultural issues that are common across many behaviors?
3. Identify at least three points when a participant could have questioned the situation to develop a better situational awareness. Discuss each of these points and how the fire service culture contributed to a lack of asking questions to develop a better situational awareness.

---

## Listen to those working in the streets

Sensitivity to operations requires that those making decisions listen to those operating at street level and that those in the street must in turn provide pertinent and accurate information to those making the decisions. Personnel operating in the street can provide a more accurate picture of what is actually happening as long as they are well trained and understand when it is important to pass information up to the decision makers. It is important to understand that decision makers can be anyone: firefighters, captains, chiefs, incident commanders, paramedics in charge, among others. Whoever those making the decisions are, they must seek out, listen for, and trust the information from those around them. As stated above, though, the information must come from well-trained personnel for the decision makers to trust that the information is accurate. For example, have you ever seen firefighters who do not always understand what is going on at the top of the department? Often, in this case, there is a lack of information that leads to rumors, misinformation, and false assumptions—all of which affect morale. Meanwhile, the top of the department may not always know what is going on in the street because each of us has a role in the department, which causes us to focus on our own specific tasks. The administration makes decisions based on the information available to them. Like a firefighter in an outlying station who may be missing pieces of information about an administrative decision, the administration may also be missing operational awareness because of their separation from the street.

> *Sensitivity to operations is about seeing what is actually happening regardless of what is supposed to be happening.*
>
> —*Dane Carley*

Let's look at a fictitious example of writing specifications for a new apparatus. In this case, administrators prefer an apparatus similar to what the department already has because it reduces maintenance costs and purchasing costs and satisfies certain Insurance Services Office (ISO) and strategic considerations. The decision about the "administrative stuff" has already been made, so the alternatives are already limited. Yet, the administrative requirements mean tall hose beds that are less functional, less interior cab space, and increased turning radius—in other words, they have an impact on the "firefighting stuff." Being sensitive to operations means the administration must listen to and consider the input from the operational staff when asking for modifications that simplify tasks, make the job safer, or generally improve company effectiveness; likewise, the operational staff must take the time to understand the strategic considerations. This way both parties find common ground to achieve all the goals as closely as possible (fig. 6–1).

**Fig. 6–1.** Illustration of the different points of view between the firefighters using the equipment and the administrators purchasing the equipment

The same processes carry over to the emergency scene. Consider the use of PPV. A decision maker orders a firefighter to turn the PPV fan into the building and begin ventilation. However, the firefighter, being sensitive to surrounding operations (awareness, training, understanding), recognizes that doing so endangers a crew searching a room near the vent point. The firefighter then communicates the important details of not being able to vent as ordered. The decision maker (listening to the one closest to the point of use), on hearing this, recognizes that the new information is important to successful operations, terminates the request, and develops an alternative solution.

Three ways to help prevent operational threats have been identified by Weick and Sutcliffe (2007, 60–61):

- Treat experience and education equally because both are equally important to success. We often place much more weight on experience because in the past there was not as much training or education specifically for emergency services.

- Do not work mindlessly or treat tasks as routine. For example, C shift does not need to dust every single Saturday even if it is not dusty, just because it is on the Saturday station cleaning list.
- Do not treat the avoidance of an incident or injury from a close call as proof of a safe operation but rather as evidence of a potential failure. For example, B shift will eventually blow something up. If no one is hurt, they will consider it safe, which is not true.

## Real-Life Example

### Event description

*Note that in this example, brackets denote reviewer de-identification.*

Engine [1, 2, 3], Rescue [1], Tower [1], Battalion [1] and Medic [1] were dispatched to a two-story residence for light smoke showing from an upstairs window. We arrived first on Engine [1] to find light smoke showing from a barely open window in a two-story four-plex of lightweight construction. Civilian accountability was attempted and all units were believed to be empty. Upon arrival of other crews with RIT lines ready, we made entry to find zero-visibility conditions and low heat. Using the thermal imager, we made our way through the first floor to the back door, which we opened for natural ventilation. The camera showed the kitchen area to be warm but not hot. In my haste I rushed the team through the downstairs area completely missing a door. We made out no victims or fire so we proceeded to the stairs behind us and went to the second floor.

Upstairs was also near zero visibility conditions with low heat. As we vented windows upstairs and eventually tried hydraulic ventilation, the door that I had missed downstairs started to burn through. The RIT team consisted of a lieutenant and one fire rescue officer who were positioned at the front door. They witnessed this progression and attempted to contact us by radio, which we failed to hear. The hydraulic ventilation obviously did not help conditions downstairs and caused most of the kitchen area to flash, pulling much of the fire upstairs. I was attempting to come down the stairs when this suddenly occurred and I was engulfed in a rollover that took up half the stairwell. I stumbled up the stairs to my team, who had no idea of what was happening just beneath us.

The RIT team immediately took action, sweeping the ceiling and pushing the fire back. The conditions improved very fast and my crew and I were able to retreat to the first floor due to the swift actions of the RIT. I hope by sharing this it doesn't happen to anyone else. It was a close call for sure and I can't thank the RIT team enough for keeping cool and taking care of business.

## Lessons learned

- My haste was the number one contributing factor in this event. If I had slowed down and "read" the environment, and not developed tunnel vision through the lens of the TIC [thermal imaging camera], we would have never missed that door.
- Make sure radios are on, loud, and locked on the right channels as soon as you get on shift.
- Venting the structure before you find the seat of the fire is OK, but I would not recommend mechanically doing so.
- Train more often with your thermal imagers.
- Train more often in zero-visibility conditions. They are becoming more and more frequent.

---

## Discussion Questions 6–3

1. Based on pieces of information in the case study about not hearing radio traffic and the recommendation to check portables, it is possible the person reporting the near-miss was on the wrong radio channel. How can the routine of morning checks lead to mindless actions contributing to near-misses at an incident later in the day?
2. The RIT acted on behalf of the initial attack group without direct orders from the attack group to do so. Discuss how this relates to the concept of comparing what is actually happening to what is supposed to be happening.

3. Thermal imagers are becoming common fire service tools. Many of us have experience using them, but few have received much education on how they work. Does this contribute to common misunderstandings or misbeliefs about what we can and cannot see in some cases?
4. How does the previous idea relate to the concept of treating experience and education equally?

---

## Adapt successfully

The first three principles focus on avoiding errors and failure. The last two principles focus on reacting to an error or failure. Weick and Sutcliffe (2007) have expressed that the hallmark of an HRO is not that it is error free but that "errors do not disable it." For the fire service, a commitment to resilience means adapting successfully by planning for the worst. It is important to be ready when an error or failure occurs, to adapt successfully. This recalls the fire service saying, "Plan for the worst while hoping for the best." Some common sayings that correspond to making assumptions and not being ready include:

- "What could possibly go wrong?"
- "Wow, I didn't see that coming."
- "We don't need mutual aid."
- "It's just a room-and-contents fire."
- "Will all the units I just canceled return to the scene?"

Adapting successfully involves three core components:

- Seeking employees with deep and varied backgrounds capable of improvising
- Developing employees through ongoing, scenario-based training
- Maintaining sufficient resources to react to unexpected events

**Improvising.** The first component is seeking employees with deep and varied backgrounds capable of improvising. Resilience is about adapting. A fire service–specific background provides the important technical skills necessary to perform the job. Nevertheless, those with varied backgrounds bring new ideas from outside the fire service that help us adapt to unexpected failures. Because it is human nature to draw solutions from previous experiences (i.e., RPDM), combining critical thinkers with a fire service–specific background and critical thinkers with different backgrounds can bring new solutions to a problem. This increases a crew or department's adaptability to inevitable problems.

**Employee development.** The second component is developing employees through ongoing, scenario-based training that challenges decision-making processes. Psychologist Gary Klein has described an interview at the National Fire Academy with a senior curriculum developer, who said, "To be a good commander, you need to have a rich fantasy life" (Klein 1998). A rich fantasy life in this context refers to imagining how a future fire might burn in a building and the various ways to attack it or imagining how an active fire is behaving currently, imagining different ways it may behave in the near future, and evaluating ways to mitigate the emergency successfully based on what is happening now and what may happen in the near future. Realistic, scenario-based training provides crews with the opportunity to experience many different situations in a controlled learning environment where they can practice and build adaptability, decision making, and situational awareness skills—as a fantasy life—to enhance their repertoire of solutions when confronted with an emergency.

**Sufficient resources.** The third component is maintaining sufficient resources on scene to react to unexpected emergencies. This is likely to be controversial because it is resource dependent. However, it also becomes an issue because of our assumptions. For example, we may assume that we do not need any more resources, we may assume we can send resources back early, or we may assume that we can handle the situation without more help. Maintaining sufficient resources reflects planning for the worst and hoping for the best. We should:

- Have a sufficient number and type of resources on an initial dispatch
- Have the resources arrive in the timeliest manner possible for your response area (e.g., automatic aid agreements)

- Ensure that those resources remain on scene throughout the entire incident. Many accidents happen as firefighters become fatigued or complacent with their increasingly familiar surroundings (e.g., calling and establishing a RIT late and sending them home early defeats the purpose if they are not there when needed).
- Ensure that the resources are rehabbed and prepared to act while on scene
- Recognize that people need sufficient rest during a shift to improve decision making

**Assessing your department's adaptability.** The easiest way to assess a department's adaptability is by looking for honest, objective answers to the following questions:

- Does the department use scenario-based training to develop crew adaptability, decision making, and situational awareness skills?
- Does the department routinely maintain sufficient resources on scene from beginning to end to adapt to unexpected events until all tactical operations are complete?
- Does the department encourage employees to seek outside resources, education, and training?
- Are employees encouraged to discuss mistakes in a near-miss reporting system or after-action review process in a learning environment?
- Are new ideas, methods, tools, and techniques encouraged?

## Real-Life Example

### Event description

It was after lunch and our station was alerted for a reported house fire. The dispatch report was fire in a second floor bathroom. The area is not serviced by a municipal water supply so tankers were added at the time of dispatch. I was the officer on the truck company and arrived behind a chief officer, who was on scene about one minute prior to our arrival. His report advised a larger single-family home with the occupant outside and nothing evident. The truck was able to position in the driveway on the Alpha/Delta corner. We met briefly with the homeowner who advised us she thought the fire was on the second floor in a bathroom. My tiller man and I entered the structure from the side Delta entrance. There was a light haze of smoke. When we entered the living room area, I made a transmission over the radio that the house was full of "junk" and for all units to use caution. The homeowner was a hoarder and there was no visible floor. Magazines and papers and boxes were all over the place.

As we proceeded to find the fire, there was still no IDLH [immediate danger to life or health]; however, the smoke was a little thicker in the living room but only about one foot from the ceiling. I heard command start placing units in service from our assignment to run another fire in a bordering company's area. I made a transmission that we definitely had a fire in this house and we haven't found it yet so don't release anybody.

My tiller man and I did a complete primary of the house and found increasing smoke conditions but no fire. It was obvious to us the fire was on the first floor. We closed windows and watched were the smoke was coming from. In the living room, the smoke increased so we focused our attention there. I have to add there were two hose lines in place at this time, one at the main entrance and another one at the side Delta entrance.

Neither hose line was advanced into the house because we didn't know where the fire was and, due to the hoarding, the thought was once you went in and got in place you were stuck. There was no room to move around or reposition a hose line.

I was standing in the middle of the living room and, using the thermal imager, started to scan the area. I saw an area of high heat on the Charlie/Bravo corner of the living room. I directed my tiller man to the area and he opened up. The area was a door and when he struck it with his pike pole, the room was free-burning. We attempted to shut the door but the fire escaped so fast we were unable to shut it. The room went from some visibility to zero visibility and high heat. We as a team recognized immediately we were in a bad situation and scrambled to exit. We were able to make it back to our initial entry point. I made a transmission to the crews at the front door where the fire was.

We were met by another crew in the kitchen. The fire was coming through the door right behind us. We were able to shut that door enough to keep the fire from entering the kitchen area. We then exited the kitchen onto the porch on the Delta side and made it to the exterior.

### Lessons learned

A hose line should have been in place prior to opening up any walls or doors when heat was present.

Be aware of your surroundings. Had we not had the opportunity to see the house with a light haze of smoke prior to the conditions deteriorating we may not have had the opportunity to escape.

---

## Discussion Questions 6–4

1. This case study gives us an opportunity to review listening to those working in the streets and transition to adapting successfully. What interaction between the company officer and the incident commander is an example of listening to those working in the streets? The company's delay in finding the fire is not a failure; however, what actions allowed the incident commander to react effectively to the fire once it was found?
2. What assumptions did the incident commander make to come to the decision to release units before finding the problem? Is this something you expect in your department?

---

## Listen to those with expertise

Weick and Sutcliffe (2007) have described deference to expertise as a habit of listening to the person with a deep understanding and firsthand, credible experience of the problem regardless of that person's rank. This is because those in managerial positions often receive filtered, politically expedient information that can lead to poor decisions.

In the fire service, this means that we should listen to those with expertise in an area that applies directly to the instance. This applies to all decision-making situations, although it may be easiest to understand in an operational example; for example, a member of the crew with electrical work experience or training (e.g., a lineman or electrician) would be a good person to consult at an electrical emergency regardless of that person's rank or seniority.

We understand that many situations in which we work do not allow for lengthy discussions, especially on an emergency scene. We are not advocating that you hold a board meeting at a fire. Rather, we are advocating that you seek input by adjusting it to the situation. An administrative situation allows for more time and ability to seek input from your people compared to an emergency scene. Nevertheless, time should be made even at an emergency scene because it is as important (if not more so) as in an administrative situation. This is most likely being done at emergency scenes already, but a conscious effort must still be made to effectively use the people on the emergency scene. Our work environment requires making decisions despite tight time constraints, high workloads, dynamic situations, and a lack of information. These factors give us a false sense that seeking out expertise is unrealistic, when in fact it is not.

If an officer knows the crew, then he or she knows their areas of expertise. Therefore, seeking out expertise should not be a time-consuming task. Officers will be able to identify quickly the person who has the best skill set to assist with the emergency and put them in a position to help solve the problem.

For example, the fire service routinely listens to those with expertise as part of normal operations when dealing with hazardous materials or medical emergencies. One reason this happens in these situations and not as often at fires is that firefighters learn early in their training to slow down and follow a methodical process at these emergencies. However, we can still access these skills during time-constrained emergency scenes (e.g., a fire), if we know our crews' skills and talents. This starts at the fire station, before an emergency, by getting to know each other and by giving all crew members encouragement to speak up later, on scene, when something important or hazardous arises. Then, we can start using it en route by asking questions while sizing up the situation.

*Organizations that live or die by their hierarchy are seldom in a position to know all they can about a problem.*

—*Weick and Sutcliffe* (2007)

Another exemplification of this behavior might be a company officer informing incident command of a unique skill one of the crew's firefighters has that would aid in resolving a situation. This provides the opportunity for the incident commander to gather specific, relevant information to improve decision making and resolve the incident more effectively—leading to increased safety and reduced property damage. In addition, this makes the crew member feel more involved while also helping the incident commander look better because it shows that he or she is a strong leader.

Pushing decision making to the lowest possible level helps the situation by gaining input from those with expertise who are performing the work. In essence, it is as though the incident commander has gained many sets of eyes and ears. This relieves the incident commander of conveying or considering the specific details of a task and provides more time for the incident commander to perform incident command tasks of organizing the scene. An incident commander is constantly trying to develop situational awareness at a scene and share that with units operating on scene to be sure units do not perform conflicting or detrimental operations. However, those performing the operations have a better understanding of the immediate problem affecting them and the possible solutions because they have the timeliest, firsthand knowledge of the situation.

Often the fundamental variable determining how (or whether) we seek this expertise is how it is communicated. Information is filtered for various reasons as it passes up the chain of command, so the message an incident commander receives may not closely resemble the original message. For example, have you ever played the telephone game and passed a message around a circle? Similarly, rank and seniority sometimes become disconnected from expertise in dealing with the specifics of an emergency. Forgoing rank and seniority, when necessary, in exchange for the expertise gained by those with specialized skills improves the probability of success because of the synergy created by exercising a collective consciousness (Ridley 2010).

## Case in Point

Apparatus respond to a report of heavy black smoke coming from an industrial building late at night. Crews have trouble finding it at first because the black smoke blends into the night sky. One of the engines finds the fire in the back of a lot. While the other engines turn around and pull into the parking lot, the first engine begins pulling an attack line to the door.

The driver of the second engine (an astute hazardous materials team member) comments to the captain that it is a metal plating shop. This means very little to the captain, so his first reaction is to not react at all. After a few seconds of thinking about it, the captain says something noncommittal but probing like, "So?" The driver explains that there are a lot of hazardous chemicals including acids in most metal plating shops. About this time, the captain on the first engine notices that the liquid pooling around the outside of the building is fuming. This combination of information is passed along to the incident commander, who then decides to halt all current operations and gain a better assessment of the situation before committing crews to an interior attack.

In the end, it is found that the pool of fuming liquid surrounding and beginning to grow beyond the south end of the building is acid. The building contained several thousand gallons of the acid in an aboveground tank that failed in the fire. The participation of the crew member with expertise applicable to this hazardous materials incident changed the outcome such that the scene was safely mitigated with no injuries.

Each of us has a role in decision making and communicating. Those performing the work maintain a high level of technical knowledge because of their daily experience on the street. Those in decision-making positions lose a lot of that technical knowledge as they take on a bigger a role in overseeing the daily operations, which takes them one step away from technical street operations. However, the decision makers gain knowledge of how all of the community's puzzle pieces fit together so they can coordinate the activities of the technically knowledgeable employees providing the services. It is important for people in each of these roles to seek input from the expert and decide when it is important to communicate relevant information. Importantly, each of us is an expert in our own role.

Pushing decision making down the chain of command and engaging those with expertise is a win-win situation. The decisions are stronger, the buy-in is higher, and the department is more successful.

## Real-Life Example

### Event description

We were dispatched to an attic fire in a single-family home. Initial arrival was an on-duty engine and ladder truck with two firefighters and one officer, who gave an on scene report of a two-story wood frame residential structure with heavy smoke showing from the rear. The officer announced command and told all traffic to go to fireground. After seeing three sides of the building, the officer (myself) ran around to back side and found heavy flames venting from the second floor gable end off the rear of the structure. At the time it was not known if this area was an attic over a first floor addition or a room on the second floor.

The officer decided a quick interior search and fire attack, pushing the fire out the already vented hole, would be the initial strategy.

The ladder truck was parked and the firefighter, officer, and an arriving part-paid firefighter donned full gear while the engine operator pulled and charged the initial attack line, set up PPV to the first floor, and gathered hand tools.

As the initial entry to the first floor was made, the next arriving on duty rescue and engine arrived on scene with two firefighters and one officer. Other part-paid firefighters were beginning to arrive on scene. The second arriving officer took command.

After searching the first floor and finding no stairs, the initial team exited the first floor and went to the rear of the structure where an exterior stairwell was found to the second floor. On initial size-up, the stairwell and two mailboxes on the house were missed, causing an approximately one-minute delay to fire attack.

Upon entry to the second floor, conditions were a light haze with complete visibility of the occupied area. Entry was made into a bedroom, which was adjacent to the first floor attic area on fire. Upon entry I noticed a mattress on the floor and a small window just above floor level.

A small pike pole was used to breach the wall while the nozzle operator stood ready and the third firefighter moved hose. After an area between the studs and about two feet tall was opened, the attack line was discharged. Conditions went from almost clear to black and steamy instantly. After spraying the nozzle for less than 30 seconds (maybe even sooner) the room became too hot to occupy. All three of us announced we had to get out, almost at the same time.

The nozzle operator and firefighter supplying hose went out the door to the bedroom. As I tried to go out the door, heat and flames prevented my departure. I felt the mattress at my knee and went for the window I knew was adjacent. I opened the window, thinking I did not want to break it because I knew others would be working under/near it.

I hung out the narrow window from my waist up. I radioed immediately asking if the other firefighters who were with me made it out, and announced I was at the window and OK. I thought to myself, "I'm OK, I just need a ladder to get down."

When I opened the window, other arriving part-paid firefighters recognized the situation was not right and immediately began bringing a ladder outside of Incident Command's direction. He was directing others in pulling secondary lines, etc. Firefighters recognized immediate action was required and took it.

While bent over the window sill, flames began coming out the top of the window. When the ladder was placed at the base of the sill I slid out head first holding the rails. The window was too narrow to attempt to turn around and the heat was building behind me. Another firefighter came up the ladder from the ground to help brace me on the descent. Once my feet cleared the window, I did an arm hook, swung my feet below me, and climbed down the ladder.

Almost the same time I reached the ground, the room and windows became fully engulfed in flames.

## Lessons learned

The greatest safety lesson that came from this was to be aware of your surroundings and maintain situational awareness. Train so that when you have to react you just do it, and do it calmly. We have done save-your-own/save-yourself training, which included quick ladder bailouts.

Take that extra time to read a building (even your average house fire where that quick initial attack will do the job). As an officer in a department with limited full-time staff, the first arriving officer has to wear many hats. Take the extra time to read the roof lines. In this case, when we opened the wall and then the nozzle, we were hitting a knee wall and not the main body of fire.

---

## Discussion Questions 6–5

What are the skills and talents of each of your crew members on your shift at your station? What types of emergencies do those skills apply to? How could their skills be applied (specifically) at those scenes? How confident are your crew members with speaking up and communicating?

---

## EIGHT BEHAVIORS FUELING MEMBERS' IMPROVEMENT

The five principles focus largely on how a department organizes its internal operations. By shaping the department's operations, these principles encourage certain employee behaviors, but the focus remains on the department. However, employees make the department's operations happen, so focusing specifically on beneficial employee behaviors helps ourselves, our department, and our profession to improve. There are specific employee behaviors that can greatly increase the chance of personal, organizational, and professional success. Knowing this, a department's goal should be to design processes that identify new employees and promotional candidates with these behaviors. Another goal should be to develop these behaviors through training and practice. All of these behaviors can be learned, but first they must be identified, taught, practiced, and, most important, encouraged.

Jeff Erickson and Lee Dyer (2004) of Cornell University have already identified eight of these behaviors. Erickson and Dyer did not study the fire service specifically; nevertheless, the behaviors they identify apply to the fire service environment and the HRO concept. That begs the question, Does the fire service have the right people in the right positions modeling the right behaviors?

For the fire service to operate as an HRO, not only do we need the right people in the right places, but we also need those people to model the right behaviors. The eight behaviors should be sought and encouraged in current and future personnel. In addition, they provide a measure by which to design education and training opportunities for our fire personnel. We need to find the right people and put them in places to model and teach the right behaviors to others. What behaviors are we trying to measure, model, and teach?

The behaviors, in Dyer and Erickson's words, are:

- Ascertain
- Communicate
- Initiate
- Deploy

- Coact
- Improvise
- Learn
- Educate

These behaviors translate readily into the familiar context of the fire service with minimal adaptation. In fact, the last three stay exactly the same. The eight fire service behaviors are:

- Be situationally aware
- Use effective communications
- Show initiative
- Be adaptable
- Work together
- Improvise
- Learn
- Educate

## Be situationally aware

It is important for all emergency responders to understand what is really happening in their environment—to be situationally aware. Emergency responders who are situationally aware have learned to constantly seek out information to better understand their surroundings. Even though it is easy to assume that we are all situationally aware, consider how many times you thought you knew exactly what was coming next only to be blindsided by the outcome. This occurs both on emergency scenes and in everyday life; for example, you may pick up a bottle thinking it is ketchup and put it on a burger only to find that it is hot sauce when you take the first bite. If you have ever experienced an unexpected event like this, then you are human and your situational awareness was incomplete. Although it is not always possible to have complete situational awareness, there are still steps we can take to improve our situational awareness both on and off the streets. We have broken situational awareness into two primary areas for simplification: operational and administrative.

Operational situational awareness means knowing where you are and what is happening during our daily activities such as fighting fires, extricating accident victims, or providing CPR. But it goes beyond knowing you are standing in front of a two-story ordinary construction building on 123 Main St. at 0200 hours. It also means understanding what is going on around you, predicting how that will affect the incident in the future, then closing the loop by making sure what was expected to happen in the future does. This means understanding and considering future consequences, such as:

- How is ICS working at this emergency scene?
- Where are your crews operating?
- Do you have sufficient resources?
- If something doesn't make sense, then why not?
- What might go wrong:
  - If you take this action, what reactions will it cause?
  - What am I missing?
- How will venting the roof affect the crew searching on the second floor?
- How does the hierarchy affect information flow up the chain of command?
- How do the department's previous experiences influence members' current decision making?

Operational situational awareness is unique for everyone because we have different experiences. Senior firefighters, because of their vast experience, have the ability to prioritize high volumes of information effectively. This helps them to create an action plan and quickly recover from unexpected events. However, this ability may also lead them to miss important details. Newer firefighters cannot process large volumes of information because they lack experience to make associations from previous incidents. Their frame of reference is based on textbooks, SOGs, and training, so they tend to see details that relate to their training (LeSage, Dyar, and Evans 2011). Therefore, newer firefighters are very valuable because they catch more of the technical details. What does this all mean? It means that by working together and encouraging information from everyone—veterans, rookies, and everyone in between—we all gain much better situational awareness because we develop a more accurate big picture (fig. 6–2).

**Novice**

- Sees more detail; misses bigger picture
- >3x the decision points
- Distractions increase decision points
- Decisions based on SOG and training
- Loss of situational awareness results in halting actions and long recovery

**Veteran**

- May overlook detail; sees bigger picture
- Decisions based on recognition-primed decision making
- Ignore or minimize distractions
- Shrug off errors and recover while maintaining control

**Fig. 6–2.** The different picture a veteran firefighter sees versus the picture a novice firefighter sees

## Case in Point

Crews arrive at a three-story wood frame apartment building with fire venting out a bedroom window into the building's eaves. The first due crew pulls a line inside to the fire apartment and begins interior operations. The battalion chief arrives and takes over command from the operating captain, making the operating captain the division supervisor. Three crews are added to the division to assist with search and rescue, fire attack, and pulling ceiling.

Meanwhile, the truck company aerial is set up over the roof with a bird's-eye view of the building. The truck captain reports that there is smoke coming from the eaves along the entire length of the building. He receives no response from the incident commander, and crews continue working as initially assigned. A few minutes later, the truck company captain reports fire coming from all vents in the roof. He again receives no response from the incident commander, and all crews continue working as initially assigned. A few more minutes pass, and the truck captain reports sagging roof trusses and implies that interior crews may be in danger. At this point, the incident commander responds to the truck captain's radio transmission with, "Copy, understand." The truck captain makes a fourth radio transmission to the incident commander stating that the attic is fully involved from one end to the other. This last radio transmission causes the incident commander to begin asking the division supervisor whether the division is making any headway. However, the division supervisor is not aware of these transmissions until his firefighter asks about the truck's reports because he had subconsciously turned his radio down earlier in the incident to focus on the tasks at hand, to reduce the distractions. The firefighter informs the division supervisor of the important messages that the truck had transmitted. At about the same time, the incident commander finally asks if the division is getting ahead of the fire because he actually heard and processed the fourth radio transmission; he is now gaining situational awareness about the extent of the fire spread.

The division supervisor's immediate mental reaction, which he keeps to himself, is, "Yes, give us a few more minutes," but he changes tack as the information is processed and before verbally replying to the incident commander with, "No, we are not getting ahead of the fire." The incident commander then orders the crews out of the building so the truck can get a defensive knockdown. As crews are exiting the building, the division supervisor exits last. The previously clear exit door on the ground floor fills with smoke as the inside environment changes rapidly because of a partial collapse. Not long after, a gust of wind blows one gable end over into the stairwell crews had been using to access the fire area.

There was critical information available to build situational awareness throughout this incident. The radio transmissions from the truck captain were clear and audible. Yet, situational awareness eluded the incident commander and the division supervisor until it was almost too late. The incident commander did not hear the first two radio transmissions because he was overwhelmed; he heard the third radio transmission but was still so busy that he was unable to process the information completely, so he replied with a reflexive transmission, "Copy, understand." The incident commander finally *heard* the fourth radio transmission, processed the importance of the message, and then took corrective action.

> While this was happening on the outside, the division supervisor inside was busy assigning crews, formulating an attack plan, and worrying that the plan was not getting ahead of the fire. The easiest way to reduce the distractions was to turn down his radio. The division supervisor did this reflexively, without any conscious thought. Had the relatively young firefighter not increased the division supervisor's situational awareness, it is likely that firefighters would have been hurt because the incident commander would not have pulled the crews out of the area in time and the division supervisor would not have agreed and reacted as effectively.

Operational situational awareness is extremely important at emergency scenes, but we often forget about administrative situational awareness. Administrative situational awareness means being aware of:

- What your emergency responders think of your department
- What citizens think of your department
- What the local council thinks of the department
- What your department can learn from other organizations in the emergency services
- How your services align with the expectations of all the aforementioned groups

The administrative aspect of situational awareness is like peeling the layers of an onion because it covers politics, budgets, grants, industry trends, human resources, the economy, planning, and much more. Just as junior and senior department members working together improve operational situational awareness, the department administration, department employees, the oversight boards and commissions, the community, and the emergency services profession must work together to improve administrative situational awareness. This means honestly evaluating the level of service and capabilities your department has or should have. For example, what is your department's ISO rating? Should it be better? Do any of the groups listed above think it should it be better? Is the cost of improving the rating balanced by the benefits? By working with the employees, community, and the industry to accurately research our situations, we gain administrative situational awareness. This helps us to develop and understand current or future plans for budgets, staffing, apparatus, stations, service levels, and more.

*One way to check your situational awareness is to see if what you think is happening and what you think is going to happen is actually what happens. If what you think is happening does not match what is actually happening, then you have lost situational awareness.*

—*Craig Nelson*

Here is how to gain operational situational awareness:

- Communicate, communicate, communicate!
- Use an aide to assist with radio transmissions and tracking resources, among other tasks
- Develop personal flags to remind you of potential hazards—situations when things are more likely to go wrong
- Set incident specific benchmarks and alternative plans to put in place if or when those benchmarks are met
- Get to know your crew
- Study, practice, and train
- Ask why (this is not the first or last time you'll see this, so it is probably important!)
- Supplement constructive criticism with possible solutions
- Get to know your equipment and response area
- Immediately advise others of hazards or potential hazards you see or hear

## Exercise 6–1

Add some of your own ideas on how to gain operational situational awareness

- 
- 
- 

Here is how to gain administrative situational awareness:

- Communicate, communicate, communicate!
- Attend community board meetings, chamber meetings, planning meetings, and neighborhood meetings
- Monitor political activity on the local, state, and federal levels because legislation being passed may affect you
- Read and write for professional publications
- Attend professional schools, conferences, and seminars
- Join professional organizations
- Ask your people about hurdles within the department that prevent them from providing quality service to the community
- Seek out information from those in the streets about their experiences in the community
- Gain Center for Public Safety Excellence accreditation
- Collect data
- Look at potential future challenges to gain situational awareness about new revenue sources, operating costs, changing demographics, and services that the community will expect
- Ask focus groups of citizens who have been educated about emergency services what services are important to them and what operating costs are acceptable to provide those services

## Exercise 6–2

Add some of your own ideas on how to gain administrative situational awareness

-
-
-

Situational awareness is critical to emergency scene management and administrative viability. The key to either form of situational awareness is realizing that we often see what we want to see and align the incoming information with our own beliefs and biases. Therefore, we must seek out challenging information from various sources and then listen to it because alternative points of view are often the most valuable.

> *If you are in a room with two or more people who disagree with you, there's a damn good chance you are the one who is wrong—especially if you're the boss.*
>
> —*Craig Nelson*

## Use strong communications

Good decisions come from good information that builds situational awareness, which is based on quality, relevant communication moving up, as well as down, the chain of command (fig. 6–3). Would you rather have a lot of relevant information on which to base decisions or would you rather make decisions using just what you see or think? This is not meant to imply that every bit of information is relevant. Relevant information is information that is directly related to the situation at the time. When it is relevant and placed into a concise format, it is intelligence. Intelligence quickly provides applicable information to those who need it.

Just as veteran and newer firefighters see different aspects of a scene while building situational awareness, they also communicate differently. For example, Klein (1998) found that experienced firefighters (those meeting certain criteria including 10 or more years of experience) have an intuitive ability to determine what information is relevant and thus should be passed up and down the chain of command.

Effective and accurate communication is imperative when working in emergency services. Emergency personnel who are good at providing intelligence make it much more likely that operations will go smoothly and that others around them will learn to communicate more effectively; consequently, good communication will become the norm. Encouraging differing viewpoints and constant communication makes us more reliable emergency responders. A firefighter, EMT, captain, or chief with a constant, diverse communication behavior builds reliability into crew operations by providing and receiving more intelligence.

**Good Communication**

- Constant communication down and up the command structure
- Teach people to speak up
- Standard terminology
- Practice communicating in scenario-based training
- Build trust and eliminate verbal attacks
- Add radio channels and aides

**Poor Communication**

- Emphasis on rank creates downward communication in the command structure
- People do not feel comfortable speaking up
- The same word or phrase means different things to different people
- Lack of trust because of previous verbal attacks
- Insufficient channels or too many channels without aides

**Fig. 6–3.** Good communication can clarify what you are seeing and increase situational awareness

This increased amount of intelligence can then be used to make decisions with better accuracy at both emergency scenes and in daily administrative operations. Fostering this environment is critical, though not always easy. It is important to:

- Teach personnel to speak up, and then encourage the behavior
- Practice talking and, more important, listening—whether in person or over the radio

- Standardize terminology
- Practice communicating during scenario-based training evolutions
- Build trust
- Eliminate verbal attacks and intimidation
- Adjust communication to the situation (emergency communication should be more frequent and concise, whereas nonemergency communication should be less often but more descriptive)
- Add radio channels to increase the availability of air space
- Delegate an aide to listen to the additional traffic and channels
- Make it personal by addressing specific people by name when the intelligence is critical
- Diminish the effect of rank and hierarchy by building a team environment; treat all members equally
- Encourage diverse, open, and honest communication

It is also important to avoid barriers such as:

- Individuals going along with the group owing to fear, laziness, or complacency. For example, en route to an emergency incident, one crew member hears a north address but does not speak up while two other crew members (including the driver) hear south, so the apparatus heads south. The one who heard north may or may not be right, but a simple question can bring all three members to the same understanding.
- Putting a positive spin on information flowing up to a superior to make the situation seem better than what it actually is. For example, we may say we are making good headway on the fire when in fact we are not.
- Not communicating often enough with intelligence. The more intense a situation is, the more often communication should be occurring. In emergency services, this is often the complete opposite of what we have been taught.

## Real-Life Example

In a well-known example from 1989, a crew of four landed a United Airlines DC10 in Sioux City, IA, without any hydraulic systems—the crew had no flight controls. Given the circumstances, they made an extremely successful landing after experiencing a catastrophic systems failure. How did they do it? Among other things, the crew did it with constant communication. The crew members communicated an average of 31 times a minute with a message each second at the peak of their communications (American Psychological Association 2014). This is one example of successful, constant communication that increased to match the increasing stress and workload.

---

The United Airlines crew communicated intelligence constantly in a time-compressed, intense, and entirely unfamiliar—and presumably impossible—situation. A plane is not supposed to have a total systems failure, and it is expected to crash if it does. For emergency services, this does not apply to radio communication alone. It also includes communications between crew members, other crews, and command staff. Each of us has a different background of experience, training, and education, so we all communicate differently. Some people are concise by nature or experience, and others are not. Those who are not concise just need training and practice. Combining this with constant communication can lead to greater success.

Making communication personal improves the reception—that is, the listening side of the equation. Instinctively, we know that addressing individuals by name grabs their attention. When you grab someone's attention right away, they are more likely to hear what you are saying because they are listening from the beginning. For example, this may be important if you ever need to call a Mayday. In fact, using a person's name is so important to critical communication that perhaps the Mayday acronym should be "NLUNAR" so it starts with the incident commander's name.

- Name (of incident commander; not just title)
- Location
- Unit

- Name (of person calling Mayday)
- Assignment given
- Resources needed for rescue

However, as humans we tend to do the opposite of this when we deliver bad news. We subconsciously, but intentionally, tend not to address the recipient of this news by name because it makes the statement less direct and easier to ignore. There are many case studies of people working in a hierarchical environment who knew they had critical information but failed to communicate it because they were afraid to confront the person in charge (Carley 2013). Some of these people even died hanging onto the very information that could have saved their own and others' lives. Therefore, if you know the intelligence you have is critical to the outcome, address the person in charge respectfully by name or title and make sure you communicate the message.

A valid argument against encouraging increased communication and adding channels at an emergency scene is the possibility of overloading the incident commander. However, there is also a solution for this situation that is often overlooked. Delegation of an aide to focus on the listening part of the equation alleviates the incident commander's workload and improves listening by adding another set of less distracted ears. Providing sufficient channels for communication and delegating an aide are strategies that foster an environment where firefighters' communications are heard and acted upon. Even though additional channels increase the workload for incident commanders, assignment of an aide counteracts this increased workload. Thus, the incident commander's ability to process the information, track the situation, and put the communicated information to use is improved.

A variation of this already used by many departments is assigning the company officer of the RIT to shadow the incident commander. This essentially makes the RIT group supervisor an aide because the supervisor is already monitoring the communications and assisting with tracking the information in case of an emergency. For those who do not have the money or the staff to add an aide position, this can be a great way to get the extra help. Plus, then the RIT group supervisor has an excellent situational awareness of the scene activities should the group be needed.

## Discussion Question 6–6

Who can be an aide, or in what other ways can command staff be increased so that having an aide becomes realistic?

- 
- 
- 

---

In contrast to the technical solutions discussed thus far, a department-level cultural solution that appears in every HRO is the value placed on diverse opinions and open communication. Seeking out and listening to diverse opinions is the single most critical component of higher-reliability communication. This type of communication is built upon trust. It cannot be turned on and off because it takes time to build.

Little things can sabotage trust built over time, so it is not plausible to treat a firefighter one way in training and expect that firefighter to behave differently on the fire ground. For example, it may be counter-productive to tell a new firefighter "You have two ears, two eyes, and two hands but only one mouth. That means you should listen, watch, and work twice as much as you speak." The intent of this advice is great—and in the right context, this is true—but unfortunately, this is often used to intimidate rather than to teach. These little things are what erode the trust that is necessary for diverse communications. Diverse communications are built upon:

- Verbally encouraging and actively protecting diverse, open, and honest communication
- Communication training beyond the typical sender-message-receiver training to understand body language and how nonverbal cues affect communication
- Encourage constant communication
- Encourage and protect diverse opinions
- Develop a trusting and open environment through strong leadership discussed in earlier chapters
- Develop emotional intelligence (EQ)

## Real-Life Example

### Event description

While operating automatic aid on a working two-alarm residential structure fire, my crew was making an interior attack in the basement of a two story house fire. There were several units advancing 1¾" handlines on the first-floor Alpha/Delta corner. My crew of four personnel encountered heavy smoke and heat conditions while advancing a 1¾" line from the Charlie side back door to locate the seat of the fire. Once the fire was located and knocked down, several attempts to notify incident command of the conditions of the assignment were made but unsuccessful due to the heavy radio traffic. This is a significant near-miss event because if a Mayday had occurred, it most likely would not have been heard due to the heavy radio traffic from the scene.

### Lessons learned

On this particular incident there were fire fighters talking on the radio with no specific task given. It was also noticed that the safety officer was giving out command orders that should have only been assigned by incident command. The radio traffic should be coming from company officers in charge of each task given and reporting to incident command only. Those that do not have a specific task should remain off the air unless there is a danger or mayday.

---

## Discussion Questions 6–7

Before posing the discussion questions, we would like to add thoughts about the incident above. This is a tough near-miss to use because we do not know the specifics. However, we would like readers to look at it in the following context: A two-alarm fire indicates several companies. It sounds like there are command positions beyond a single incident commander (e.g., safety officer, operations chief, group/division supervisors).

1. Consider whether there was too much communication or whether there were too few channels for communicating.

What is your crew's opinion and why? Does this example seem like the way to handle this based on this section's discussion?

2. Should the only time a firefighter talks on the radio be for Maydays and emergencies? Why or why not?
3. How is command supposed to gather information to make decisions without information traveling up the chain of command?
4. How does or could your department tweak your response to add a command aide without affecting operations significantly?

---

## Show initiative

Showing initiative is where action taken of one's own volition will improve the outcome or status of the current situation. This can be as simple as seeing that the truck is dirty and washing it or seeing a piece of garbage on the floor and picking it up. But you would never get in trouble for washing a truck or picking up trash.

The level of initiative we strive for requires more guts because it may mean taking action or speaking up when there is no clear-cut response or action to take. True initiative entails acting outside your department's SOGs but within your ability to improve the situation. We are not talking about situations where you go against established SOGs or practices. Rather, we are referring to situations where there is no SOG or where there is an SOG but it does not appropriately cover the situation. Many of the situations we encounter in the streets are not black and white; they are shades of gray but must still be resolved. When we encounter these situations, we have to make decisions about whether to act and how to act if we do. The overall intent of showing initiative is to improve the situation. A starting point for taking initiative is determined by answering:

- Is the intent of my actions to improve the situation?
- Can I explain to supervisors and customers why I took those actions?
- Is the method in which I implemented those actions justifiable?

- Would I be able to explain my actions knowing it is most likely being filmed by the public?

### Case in Point

Crews are working at a three-alarm apartment fire with fire on the top floor and attic space of a three-story wood frame apartment building (lightweight construction with a truss roof). Crews aggressively attack the fire within a strong ICS organization that develops great situational awareness among the crews operating on scene. Suddenly, the roof partially collapses, and a firefighter issues a Mayday because the collapse cuts off his retreat. He appears at a third-floor window. A crew on the ground, without an order from command, brings a ladder to his window and removes him to safety.

## Discussion Questions 6–8

1. Is this an example of freelancing, heroism, or showing initiative? Is it freelancing if the situation is not a Mayday or ends unsuccessfully?
2. What would you have done in this situation? Does it change your answer as to how you would label it?
3. How would you keep this from being freelancing?

---

We often label emergency responders who were in the right place at the right time heroes or great firefighters, but those who were in the wrong place at the wrong time are often labeled as poor firefighters or poor EMTs. We are not leaving good firefighting or EMS to luck or implying that intelligence, skills, experience, and training do not make for better emergency responders; rather, we are just saying that sometimes the outcome of a situation is beyond our control, and luck can influence how the situation is viewed in the future. This stems from a societal norm dating back to the 1400s of framing our perception of an action's merits on the outcome (Dekker 2011). This norm causes us to judge someone's actions without even realizing the outcome is influencing our opinion of their actions. Thus, perception invariably frames our judgment. One of George Carlin's jokes sums this idea up well. To paraphrase, he asks if you have ever noticed that every driver who passes you is a crazy maniac, and every car you pass is a hazard and should not be on the road.

> *Studies have shown that 90% of error in thinking is due to error in perception.*
>
> —*Edward de Bono*

When used appropriately, ICS allows responders to show initiative at an emergency. This is the difference between freelancing and showing initiative. If responders are properly reacting to a situation and showing initiative, they *must* communicate their actions and intelligence to command. By contrast, a freelancer would react without communicating what is happening or what is being done. Responders reacting to a situation by showing initiative do so appropriately and safely by communicating their actions up the chain of command to maintain everyone's situational awareness.

## Real-Life Example

### Event description

At 2106 hours [Engine 1] was dispatched to a fast food restaurant for a fall injury. [Engine 1] found the patient, who was an employee, at the top of a stairwell that leads to the basement storage area. [Engine 1] began the regular line of questioning and treatment for what seemed to be a standard medical call. The patient was a 24-year-old female who was pregnant. The patient stated that she was going into the basement to check on something and became lightheaded and fell. One of the other employees heard the fall and went to the stairwell to help the patient. Both employees exited the stairwell and called 911 to report the fall/injury. As the Captain from [Engine 1] was questioning the patient and one FF [firefighter] was checking vitals, the other FF and the Engineer went into the basement to see if the patient had tripped or slipped on something. Shortly after entering the basement both members of [Engine 1] became lightheaded and exited the basement. Upon exiting the basement, the Engineer fell and both members reported dizziness and a bitter taste in their mouths. [Engine 1]'s Captain immediately

- Promoting from within. In a bit of irony with respect to the last point, promoting from within contributes to organizational memory and skill-specific experience. Although it is important to look outside to bring new ideas, it is also important to promote from within to maintain a level of stability. Hiring, even occasionally promoting, from outside combined with primarily promoting from within for stability blends diverse backgrounds, new ideas, and skill-specific experience into the culture. Communication is the delicate mediator between diversity and stability (fig. 6–4).

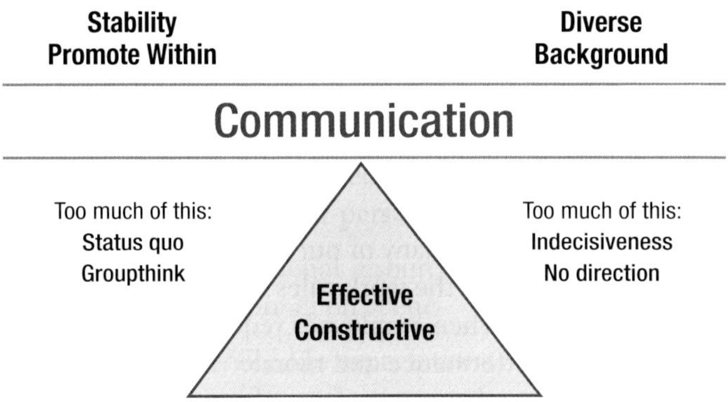


**Fig. 6–4.** The balance between diversity and stability in an organization

- Stability. It is important to develop a cohesive group, which depends on group stability. This leads to trust, which enhances communication (Carley 2013).

- Encouraging ongoing training and education. Adaptability is already an aspect of fire service culture. This cultural trait is passed to younger firefighters through ongoing training and education. Therefore, it is important to engage all members in this activity so that the senior firefighters can pass this trait down through their experiences and stories. More knowledge provides adaptability—essentially more tools in the bag of tricks. This resembles driving: When you drive to a city 500 miles away every couple of months, you know other routes exist but really know only one route well; however, you probably know five routes to the grocery store 5 miles away and adapt to weather, time of day, traffic, whether school is in session, and other changing conditions.

## Discussion Questions 6–10

1. What are some examples you have experienced of adapting (administratively and operationally)?
2. Is this behavior being passed on in your department? How is this behavior passed on in your department?
3. Which does your department lack—stability or diverse backgrounds?

---

## Real-Life Example

### Event description

*Note that in this example, brackets denote reviewer de-identification.*

We were dispatched to a car fire at a busy intersection. Upon our arrival, the first due engine was on scene and the crew was pulling hose and their [extrication] tool. The crew went to cut the car hood and the [extrication] tool wouldn't operate properly. On top of that, we had a new engineer driving the engine and he was struggling to get water. At this point, my crew was dismounting the apparatus. By now, the crew was trying to open the hood with pry bars as water was being applied and command established. The first due crew and my crew were trying to open the hood of the car and it was taking too much time as the fire was progressing. We got the hood opened enough that we could see the locking device. I went and got bolt cutters and cut the U-Ring and we got the hood opened and the fire was extinguished.

### Lessons learned

The generator for the [extrication tool] was not working and the engineer never realized it. This is a communication issue. Know the equipment. Teamwork: work more efficiently as a team. Training: adapt to conditions when things don't go your way.

---

### Case in Point

Many departments are changing their tactics at active-shooter incidents based on emerging data. Emerging data shows that providing EMS care as quickly as possible is critical to victim survival. This has led to EMS personnel operating with law enforcement protection in warm zones of a building that still contains an active shooter in the hot zone. The development of these strategies cannot be done by the fire department, EMS provider, or law enforcement alone. It is critical that all parties who hold an interest in such an operation are involved from the ground level in developing the strategies and tactics that will be used. After the strategies and tactics are developed, training with all of the players is critical to not only improving the plan but getting to know the nuances of each player's operations. This type of situation has proven to also improve relationships and communication at all incidents as the trust and understanding of complimentary agencies grows through practice and increased interactions at a one-on-one level.

Imagine the difference between a high-performance, cohesive fire crew and a new crew or a crew who are unfamiliar with each other. We understand the concept of cohesive crews, but we sometimes overlook opportunities to build similar relationships with outside agencies with whom we normally work closely at emergency operations.

How would the following improve our relationship and, by extension, our operations with other agencies?

- Developing working SOGs that complement each other's processes
- Getting to know each other on a personal level before the event
- Training together regularly to reinforce the first two steps
- The fire department trains law enforcement on how to operate safely at a fire scene, and law enforcement trains firefighters about how to operate safely at a crime scene.

We do these same things to build cohesiveness in fire crews to bring operations to a higher level and provide better service to the public. It is important to remember that high performance builds cohesiveness, instead of the converse; hence, practicing together regularly builds cohesiveness (Carley 2010). When this can be accomplished, we are making operations safer not only for others but also for ourselves. Picture mutual aid agreements with some basic response planning—but not just with other fire departments.

Engaging others is a fundamental aspect of people working together. This helps avoid groupthink and brings new ideas to complex and difficult emergency operations.

## Real-Life Example

### Event description

The three-person EMS crew responded to a call dispatched out as medical unknown problem, patient in pain. The patient disconnected before any more information could be obtained. Upon arrival at the mobile home, the police arrived and approached the home with weapons drawn. One EMS person at the front door remained with the officer until his partner motioned for him to return to the ambulance. Once the police officer cleared the mobile home and found a subject possibly dead from a self-inflicted gunshot to the head, the officer motioned for the EMS crew to come in. He did not secure the handgun next to the victim. A request was made to secure it, but the police officer stated it would interfere with the investigation.

It was eventually secured, and the crew went in and confirmed death with the medical control approval. After investigating the call with dispatch and listening to the recorded 911 tape, it was determined dispatch did not recognize what a "45 lock and loaded" meant. They thought it meant a loaded gun, but locked in a safe. This information was never relayed to the EMS crew, but was given to the police officers.

### Lessons learned

Better communication would have prevented our crew from even approaching the scene until police officers cleared it. It is also important to note that our EMS crews should have pulled back the ambulance to a safe distance as soon as the weapons were drawn by the police officers. Our fire department has since implemented training with dispatch and police on proper use of priority dispatch for every call involving police. Police were refusing to follow priority dispatch due to time involved.

## Discussion Questions 6–12

1. Information was available to improve the crew's situational awareness (e.g., the officers' drawn weapons). How could they have collaborated to improve the situation ahead of time?

2. Who could your department collaborate with?

---

### Improvise

Improvising is about not mindlessly following a policy book. People in emergency services often encounter rapidly evolving situations, and responders adapt through improvisation. The improvisation behavior is learned from and encouraged by an organization emphasizing value-based decision making. This gives flexibility to emergency responders, who often encounter situations where there is no script. By contrast, a rule-based organization tries to create rules anticipating every possible situation that may be encountered. The rules tend to tie people's hands or restrict their ability to improvise (fig. 6–5).

**Fig. 6–5.** Too many rules can actually restrict the number of options available for solving problems

## Case in Point

Two firefighters are injured in two different fire departments. The firefighter's administrations each meets separately and end up addressing the events leading up to those injuries in very different ways.

The first fire department reviews policies and procedures in place to prevent such incidents. Not surprisingly, they decide that they need to add a new rule to the current list of policies and procedures to cover the new variation of the incident. This attempt to address the recent situation or problem is now resolved in this administration's mind. The entire agency can read the new policy or guideline and know never again to do exactly what happened in the inciting incident. If the same pattern of incident and response does happen again, someone can be punished quickly for breaking the newest policy or guideline in the already encyclopedia–sized SOG set. They move on feeling good that the situation or problem was "fixed" quickly and is now in the rearview mirror.

The second fire department decides to approach the situation or problem by attempting to find the true cause(s) of the incident. They plan to take this information and produce proactive training evolutions to prevent this pattern of incident and response from happening again. This philosophy recognizes two axioms. First, the department expects employees to make decisions based on the department's values —that is, show initiative. Second, decision makers are more effective when following commonsense guidelines based on an established fire department philosophy than when trying to remember a 1,000 page rule book when operating under pressure. In this environment, decisions are wrong only when they are counter to the department's values and operational philosophies, which naturally encompass firefighter safety. The employees are not expected to remember an endless list of rules, but rather use their good judgment instilled by the department's training and culture. Personnel rely on training, experience, and common sense when they encounter a situation that could lead to a problem. The department expects crews to base their decisions on the department's established service philosophy because guidelines and policies cannot possibly be written for every scenario they will encounter.

The first fire department in the example above encourages rule-based decision making. The second department is an example of encouraging value-based decision making. Comparison of the two highlights their different advantages and disadvantages:

- Rule based ("The Quick Fix"):
  - Pros: Quick; easy; widely used (because it is easy); provides finality
  - Cons: Does not completely solve the problem or find the root cause, only masks it while the root cause continues to develop into a larger, more serious problem; leads to a list of SOGs so long that no human could possibly remember them all (and how effective are SOGs that cannot be remembered?); is reactionary ("knee-jerk"); often singles out individuals; most important, employees lose their flexibility to improvise because they are not allowed to make decisions
- Value based ("The Obi-Wan Kenobi"):
  - Pros: Proactive; finds the true root cause; no memorization of an unrealistically long list of SOGs; teaches crews to approach problems as a team; most important, it is proven to be effective
  - Cons: More difficult; a department is never done solving problems (i.e., a work in progress); takes longer

Not all guidelines and policies are bad. In fact, we need some policies and guidelines to set clear expectations and establish the department philosophy that will guide people in their decision making. Employees must first be empowered to make the decisions for a department to see the benefits from improvising. This employee behavior can be developed only in an organization that wants to learn and provides an environment in which that can happen. However, this represents a paradigm shift for most departments because our operations have been rule based for so long. Because this is all that we are familiar with, we often have a difficult time seeing and accepting other methods even when they have been proven more effective.

> *Individual commitment to a group effort—that is what makes a team work, a company work, a society work, a civilization work.*
> —*Vince Lombardi*

What types of behaviors help to develop a culture that accepts improvising? The culture is developed by several key components including:

- Respecting the validity of every firefighter's idea regardless of rank or seniority
- Setting an example as an officer of wanting to learn by admitting mistakes and explaining *why* they are mistakes (how do they diverge from the department's values, philosophy, and mission?)
- Providing communication tools to enhance people's ability to contribute to the organization
- Reacting to a situation or problem pragmatically, logically, and calmly instead of yelling, screaming, or instantly writing a new rule
- Providing employees with the training they need to understand the department's values, philosophy of operations, and mission (which assumes that the administration has taken the time to write these documents)
- Most important, except for basic game plans for initial fire attack or other emergency operations, writing SOGs that assume employees are capable (because they are) of making logical decisions based on the department's values, philosophy of operations, and mission

In themselves, these basic steps do not build a culture; rather, they are ways to enable building the necessary culture. Mark Wallace does an excellent job of diving into the topic of building a values based organization in detail in *Fire Department Strategic Planning: Creating Future Excellence* (2006).

The following case study is a good example of a situation where there was not a guideline or policy to address the specific situation. It summarizes an incident where a guideline or policy may not be necessary or may not have done any good to help in this situation and is an example of how our people need to be able to improvise on a daily basis.

## Real-Life Example

### Event description

Our department responded with police for a 911 call by the homeowner for a working structure fire (fire in the basement). Upon arrival at the scene, police found a two-year-old child outside the front door and removed her from the scene. Smoke and fire were showing when the scene size-up was made. When the initial crews were setting up initial operations, gunfire was heard. Our first-in engine was hit with gun fire four times and several of our members were in the line of fire. Fortunately no one was injured in this event. Once the gunfire ceased, entry was made into the residence and the fire was extinguished. Three bodies were then recovered. The scene has been ruled a murder/suicide and the fire was intentionally set by the subject firing upon our crews. In retrospect, this was an ordinary scene at the onset. Reports from the initial crews on the scene show nothing to raise their suspicions about the events that took place. Our initial operations were being performed under the direction of incident command. Our initial response to this fire was 3 engine companies, 1 truck company, 2 ALS ambulances, and 1 Chief. Upon arrival and after the initial events, the response was upgraded to using 3 mutual aid engines to support our stretched resources, another Chief, and a call back of employees.

### Lessons learned

The lessons we learned are again that every scene must be treated in accordance to current policies, but still understanding that any scene can change at an instant. We have no suggestions to prevent a similar attack as this was a very unusual situation. Actions used to correct the situation are not something that we are really trained to do. We are not supposed to secure scenes of violence. Again, without any clues this was going to unfold in our crews being shot at, there is nothing else we could do. We have no SOG regarding this situation specifically. The only policy on [this] we have is to have the scene secured prior to our entry. Unfortunately, police and fire arrived simultaneously, and then the gun fire began. Our members did their best to cover and conceal behind their apparatus.

## Discussion Questions 6–13

1. Should this department have or create an SOG for this situation?
2. Can or should an SOG be written for every situation? Why?
3. Broken rules often result in discipline. Does this solve the problem? Why?
4. What would a value-driven philosophy solution be for this situation?

---

## Learn

Learning is about mastering a skill so well that it becomes second nature. The learn behavior prevents us from failing when under pressure at emergencies.

### Case in Point

Often, because of tradition, new firefighters begin their career in a fire academy or some other type of organized training by practicing the same skills over and over and over. We pull hose, bed hose, pull hose, bed hose, pull hose, bed it, and then pull it again. We throw a ladder, take it down, throw it again, take it down, and throw it again. We practice each critical firefighting skill until it becomes second nature. The combination of being young and feeling confident in our skills leads many to exude this confidence toward the end of the training—some may even become cocky. After graduation, the new firefighters move to a fire station and begin responding to calls. In their mind, life is going to be easy because all of the training is done. Right? It is likely, however, that the new firefighters will find company officers who continue drilling the skills (hopefully). It is also likely that senior firefighters will knock them back a couple of notches, which when done appropriately can help the newer firefighter, but if hazing is used then it loses its effectiveness. It seems like some of our senior personnel enjoy reminding these "kids" that they are lower than pond scum.

The repetition described in the process actually fits the learning behavior. However, knocking the kid down a few notches is a behavior that typically fits the definition of hazing, which is inappropriate and counterproductive behavior in a learning culture.

A constant drive to learn as a firefighter, company, shift, and department is a fundamental behavior for safety, improvement, and striving for excellence. The fire service has a history of testing for and hiring the most qualified individuals. However, learning about the profession should not stop until the day of retirement—and for those who exhibit the learning behavior in the fire service, it truly does not.

*Winning doesn't come from want, it comes from preparation.*

*—Mark vonAppen of Fully Involved*

## Educate

People who exhibit the educating behavior seek out knowledge and share it. This knowledge often comes from within the profession in which they work, but some of it must come from outside. Some of the most innovative ideas come from other professions. People who exhibit the educate behavior are the ones who see, read or experience something; they stop what they are doing, seek to understand it, and then share with others why it is important to themselves, their department, and the profession (fig. 6–6).

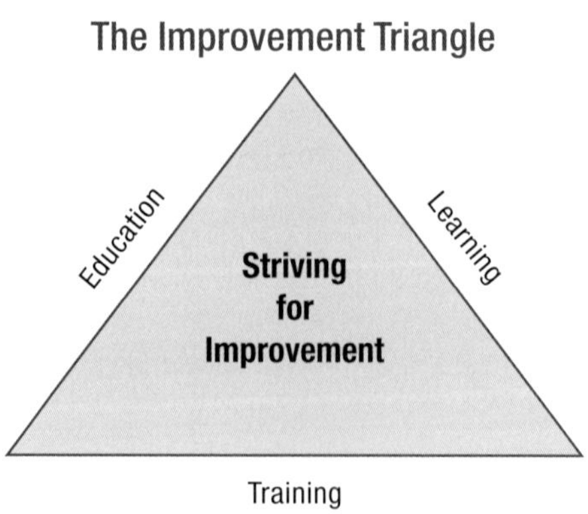


**Fig. 6–6.** There is more than one method for self-improvement. Using only one of the methods is an incomplete approach.

## Case in Point

One officer or firefighter organizes or participates in a couple of hours of training every day on which the crew works; this hones the crew's skills at firefighting, EMS, hazardous materials, and technical rescue, among other skills. A second officer or firefighter sits down with the crew every workday to review internal lessons learned or a national near miss report; this crew learns what works and what does not from others' successes and failures. A third officer or firefighter attends the National Fire Academy; this officer or firefighter learns ways to improve the department's operations on the basis of fire service research. A fourth officer or firefighter goes to college to obtain a bachelor's degree; college exposes this person to ideas and potential solutions (maybe with a little tweaking) from other industries.

Which of these four individuals is doing the most good? What is the difference in their approaches to improvement? There is no doubt each of these individuals is learning things critical to the success of the fire service, but is there much difference in what they are all doing? Are these examples of education, or are they examples of training?

The fire service often blurs the distinction between these methods of learning, but they are very different. It is necessary to understand what each achieves so that we can correctly apply that type of learning to a given situation in your fire department (fig. 6–7).

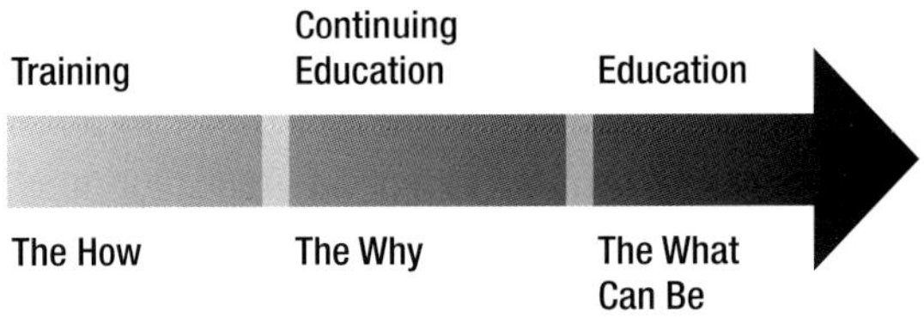


**Fig. 6–7.** Illustration of the different thought processes gained through each method of self-improvement

All these methods of learning have different definitions, intents, and learning outcomes:

- Training. This teaches or reinforces fundamental skills ranging from something basic (e.g., donning an SCBA) to something more advanced (e.g., incident command decision making). Examples include:

  - Fire ground training
  - ICS simulations
  - Pump operations
  - Apparatus driving
- Continuing education. This is accredited training and curriculum to maintain skills and abilities to a tested standard. Examples include:
  - College certificate programs
  - Qualified or tested training hours used toward professional recertification (e.g., EMTs, teachers, law enforcement, doctors, pilots)
  - Certain National Fire Academy classes or programs (e.g., command and control, fire investigation, Executive Fire Officer Program)
- Education. This comes from an accredited school and provides a broad knowledge base of critical thinking skills while promoting reliance on research from numerous sources. This may be:
  - Earning a high school diploma
  - Earning an associate's degree
  - Earning a bachelor's degree
  - Earning a master's degree
  - Earning a doctorate (PhD)

*Experience teaches someone how to do something without necessarily teaching why it is important, while education teaches why something is important without necessarily teaching how to do it. That is why both are important.*

*—Dane Carley*

To fully understand the educating behavior and get the most from it, we must also understand the different ways to achieve it. Each of the above methods reinforce a different aspect of the educate behavior and are important to the overall improvement of ourselves, our department, and our profession.

## BRINGING IT ALL TOGETHER

As illustrated in figure 6–8, an HRO is a large system comprised of many parts that support each other much like the fire tetrahedron. However, when broken into smaller pieces, it is a manageable structure to help develop a more successful department.

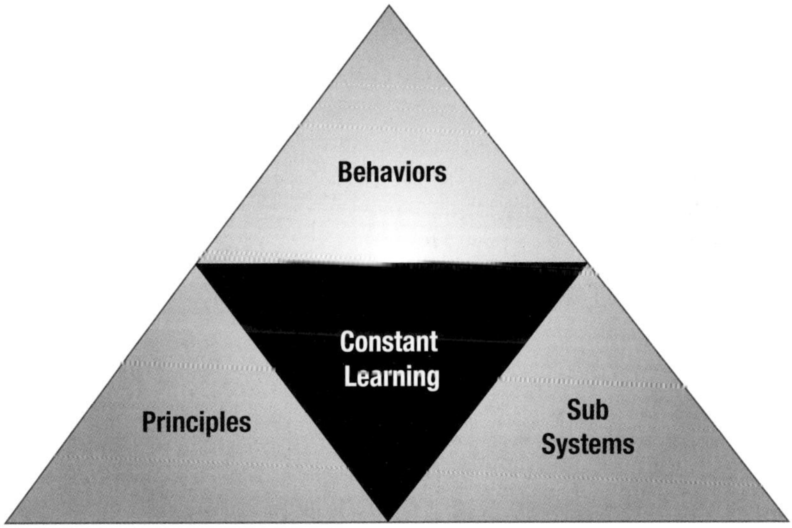


**Fig. 6–8.** The three major parts of an HRO and how they interact to support the learning culture

The cornerstone of an HRO is constant learning, much like the chemical reaction in the fire tetrahedron. Constant learning helps the department as a whole. The three legs of a tetrahedron are the principles, behaviors, and subsystems. The five principles support the department's work to ensure that it is successful. However, departments do not exist—and cannot be successful—without the people in the department. The eight behaviors help the people in the department to be successful so that the department can be successful. There are also subsystems that have HRO traits (fig. 6–9). The fire service already uses versions of these systems. The ICS, particularly FIRESCOPE ICS, and CRM are two examples of subsystems with HRO traits.

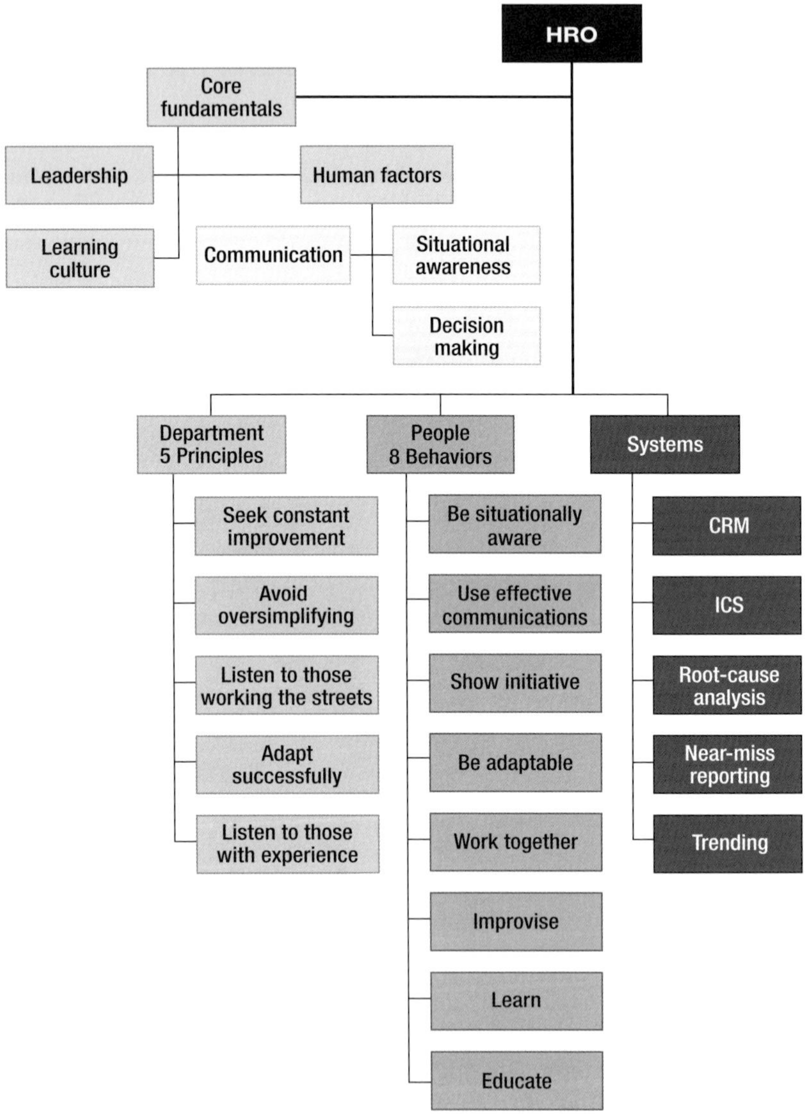


**Fig. 6–9.** How all the pieces of an HRO fit together

The single most important action you can take in an HRO is to teach yourself and the people in your department to ask one question: Why? Asking why as a department supports the five principles, and asking why as an individual supports the eight behaviors. Both support the learning culture. All of this helps a department to strive for success.

## REFERENCES

American Psychological Association. 2014. *Making Air Travel Safer through Crew Resource Management*. American Psychological Association website. http://www.apa.org/research/action/crew.aspx.

Carley, D. A. 2010. *Organizational and Group Influences on an Individual in a Fire Department Company*. Fargo, ND: Fargo Fire Department.

———. 2013. *Communication Impediments and Support in the Fargo Fire Department's Organizational Learning Process*. Emmitsburg, MD: National Fire Academy.

Dekker, S. 2011. *Drift into Failure: From Hunting Broken Components to Understanding Complex Systems*. Burlington, VT: Ashgate Publishing.

Driessen, J., L. Outka-Perkins, and L. Anderson. 2005. *Two Entrapment Avoidance Projects: Studying Crew Cohesion as a Social Human Factor*. Missoula, MT: USDA Forest Service, Missoula Technology & Development Center.

Erickson, D., and L. Dyer. 2004. *Toward a Strategic Human Resource Management Model of High Reliability Organization Performance*. Cornell University School of Industrial and Labor Relations website. http://digitalcommons.ilr.cornell.edu/cgi/viewcontent.cgi?article=1008&context=cahrswp.

International Association of Fire Chiefs. 2003. *Crew Resource Management: A Positive Change for the Fire Service*, 3rd ed. http://www.iafc.org/files/1SAFEhealthSHS/pubs_CRMmanual.pdf.

Klein, G. A. 1998. *Sources of Power: How People Make Decisions*. Cambridge, MA: MIT Press.

LeSage, P., J. T. Dyar, and B. Evans. 2011. *Crew Resource Management: Principles and Practices*. Sudbury, MA: Jones and Bartlett.

National Fire Fighter Near-Miss Reporting System website. http://www.firefighternearmiss.com/.

National Park Service. 2001. *Cerro Grande Prescribed Fire: Board of Inquiry Final Report*, February 26. http://www.fireleadership.gov/toolbox/staffride/downloads/lsr6/lsr6_Cerro_Grande_Board_of_Inquiry_Report_Feb_2001.pdf.

Ridley, M. 2010. *The Rational Optimist: How Prosperity Evolves*. New York: HarperCollins.

Wallace, M. 2006. *Fire Department Strategic Planning: Creating Future Excellence*. 2nd ed. Tulsa, OK: PennWell.

Weick, K. E. 1990. "The Vulnerable System: An Analysis of the Tenerife Air Disaster." *Journal of Management* 16 (3): 571–593.

Weick, K. E., and K. M. Sutcliffe. 2007. *Managing the Unexpected: Resilient Performance in an Age of Uncertainty*. 2nd ed. San Francisco: Jossey-Bass.

# 7

# INFORMATION MUST FLOW UPHILL TO MAKE GOOD DECISIONS

## CHAPTER OBJECTIVES

- Reestablish the incident command system (ICS) as a communication, situational awareness, and accountability tool
- Describe how key ICS components support a higher-reliability organization

# INCIDENT COMMAND IS MORE THAN BOXES TO CHECK

The first introduction to the ICS for many of us is during Firefighter I and II training or during departmental training as part of a federal class such as NIMS ICS-100. One feature the Firefighter I and II classes and the federal classes have in common is the description of ICS as an incident management tool (International Association of Fire Chiefs 2010; *Fundamentals of Fire Fighter Skills* 2009; Stowell and Murnane 2013). As a management tool, there is often little focus on true ICS capabilities beyond checking the boxes of the ICS organization. ICS is often taught to us as a black-and-white management tool. However, the more important feature of ICS is more of an art than a science, and that is the leadership aspect. Much like we know that there is a science to fire behavior, it is the combination of the science and art of firefighting, beyond merely applying water to the fire, that makes us truly effective.

ICS is a great management tool—and much more. The management aspect of ICS provides a division of labor to accomplish seemingly overwhelming feats and provides safety through accountability. Although ICS may be viewed as a management tool—and thus primarily an administrative tool—ICS is in fact an effective tool for everyone on scene, and its greatest potential is as a communication tool that can contribute to situational awareness for everyone operating at a scene. What can ICS provide? Through the many ICS tools used as intended, ICS provides:

- Comprehensive communications that can flow up and across the chain of command in a common language
- A shared understanding, or "common operating picture" (Deal 2010), of the situation at hand
- Improved situational awareness through effective communications
- Accountability through an effective, manageable span of control
- Division of labor to prevent overwhelming any single resource, so decision making is improved (Bigley and Roberts 2001)

- The expectation that the best expert, instead of simply the most senior or highest-ranking person, is a part of the decision-making process
- A formal after-action review that enhances the department's ability to learn from an incident

These ICS traits relate directly to the HRO concept. Comparing these traits against the five principles and eight behaviors reveals that the ICS is an HRO system when used properly, rather than just a management tool.

## Comprehensive communications, situational awareness, and shared understanding

The fatal flaw with viewing ICS solely as an incident management tool is that it implies a hierarchy. Because of the fire service's customs and historical operation within a hierarchy, this implies that communication should more often than not coincide with decision making at the top. Hence, the flow of communication is generally from top to bottom as orders are issued when the incident commander makes decisions.

However, ICS is primarily a communication tool, and ICS works only with "intense communicating" (Bigley and Roberts 2001). ICS is designed to develop situational awareness, which in turn leads to a shared understanding by communicating key pieces of information consistently. Beyond the ICS boxes, what are the tools ICS provides?

- An incident briefing with an area to sketch the incident, so everyone can visualize the same scope and arrangement of the scene
- The incident objectives, so everyone is trying to achieve the same outcome
- The planned strategies and tactics, to avoid duplication of effort or, worse, conflicting efforts that may lead to injuries
- An understanding of what is going on nearby, so a crew can compensate for changing conditions without endangering those around them

- Clear and specific work assignments for resources operating at the scene and an understanding of how those assignments affect others operating on the scene
- Ongoing status updates

Although these tools may seem cumbersome to use on a small incident, they are the most valuable tools on any incident regardless of size. This information must be communicated to develop a shared understanding from the incident commander to every single resource on an incident. Why is this so important? Besides the obvious reasons such as avoiding duplication of work or conflicting work, this information puts everyone on scene in a position to make competent split-second decisions to accommodate changing or unexpected conditions because all personnel will know what is happening across the incident.

### Case in Point

Units are responding to a fire in the first-floor kitchen of a two-story, single-family dwelling at 0200 hours. The first unit arrives and establishes incident command. The first unit suspects family members are still in the upstairs bedrooms because of the time of day and because no one met them in the front yard. The initial incident commander feels that the best way to minimize the family's risk is to confine and extinguish the fire first because other units will be on scene very shortly to conduct search and rescue and ventilation. Therefore, the three-person crew pulls a preconnect toward the fire in the kitchen in the back of the first floor.

The battalion chief shows up and assumes incident command from the initial company officer. The other units are beginning to arrive. The new incident commander communicates to the truck company that the truck's objective is to accomplish ventilation in support of interior operations that include confining and extinguishing fire and search and rescue of the second floor. The second engine is assigned search and rescue of the second-floor bedrooms after being advised that the first engine has a line on the first floor and the truck is performing ventilation. The third engine arrives and is assigned a backup line at the bottom of the stairs to the second floor after being advised of the other companies' activities and locations.

As the truck crew is opening the roof on the old house, they encounter heavy fire. The kitchen fire has penetrated the walls of the balloon-frame home and spread vertically. The truck company advises the incident commander, and the other operating companies hear the radio traffic. Before the incident commander has time to adjust, conditions deteriorate rapidly on the second floor, where no victims have been found yet. The search crew is reporting deteriorating conditions and that they need assistance.

Because the backup company and the truck company understand the situation, the backup hose team is able to first tell the company on the first floor that they are moving upstairs and then move their line into position at the top of the stairs to protect the searching crew. Meanwhile, the truck company is able to quickly adjust its operations and reposition ladders near the upstairs windows in case the search crew needs to bail out.

This may sound like freelancing. And it would be if companies were not reporting their actions to the incident commander as they adjusted to the situation. However, it is not freelancing because the companies communicated their actions up the chain of command. It is possible for companies to predict these needs because the basic components of the ICS created a shared understanding of what was supposed to happen, what is happening, and what needs to be done to compensate. Because the units are reporting their actions up the chain of command and maintaining effective communication, the incident commander is more aware of the changing conditions, able to maintain accountability, and compensate by assigning another company to back up the crew on the first floor.

Obviously, this is a relatively straightforward example, but it nevertheless illustrates the most basic function of ICS. The incident commander, by developing a shared understanding that places everyone on scene in the same play, created a situation where crews could anticipate and complete the adjustments to overcome changing conditions without waiting for orders from the incident commander. The crews were able to adjust and communicate that information to the incident commander because they were aware of the situation and understood how the changes affected other crews. The communication up the chain of command gave the incident commander a better understanding of the current situation and the ability to plug holes created by the crews' adjustments. This example illustrates that communicating up the chain of command is as important—and in some cases even more important—than communicating down.

The crews' ability to perform tasks to accommodate conditions is based on the incident commander's ability to communicate the objectives and the complementary tasks that need to be coordinated. An incident commander may tell an experienced truck captain, "I need this fire ventilated," after conveying the objectives and tasks being completed by other crews on scene without any further guidance. This leaves the truck captain the option to do what he or she thinks is best for the situation. However, the same incident commander may instruct the same truck captain to employ very specific ventilation tactics at the next fire because of a specific concern at that fire, such as either a hazard that was not present at the previous fire or, more generally, anything that warrants particularly close coordination. However, the truck captain should still have the option to adjust as necessary if the conditions change, but these adjustments would need to be communicated up the chain of command. This is not freelancing if the adjustment supports the operational objectives, does not threaten the other crews on the scene, and is communicated to the incident commander.

This works best, however, when one person in the crew is able to supervise without being heavily involved in the actual task. It takes an awareness of surrounding conditions to adjust to changing conditions. The more involved a supervisor is with the task, the less ability the supervisor has to recognize and comprehend the changing conditions and how they affect the overall situation.

The incident commander cannot make solid decisions to coordinate actions without information flowing up the chain of command. ICS is designed to be a two-way communication tool, yet the fire service culture tends to discourage upward communication and emphasize downward communication. Crews and supervisors should both be communicating their surroundings and actions up the chain of command constantly so that the incident commander has a clearer picture of the entire situation. An example of this is giving a CAN (conditions, actions, needs) report at regular intervals or when conditions are changing.

## Span of control improves decision making

A person's brain, like a computer, has a finite amount of processing ability for decision making. Unlike computers, however, people do not have a rating that tells coworkers their capacity for processing information. This is because a person's ability depends on many circumstances specific to a particular situation, such as how well rested the person is, the presence of stressors from outside work or stressors specific to the situation, previous experience with a similar situation, and an inclination to seek additional resources or help all contribute to a person's ability to process information, understand a situation, and by extension, make decisions. As each one of these circumstances compounds negatively, the resultant increased stress affects the person's ability to process information and make good decisions (fig. 7–1). The key here is that stress contributes to bad decisions by reducing our ability to (Klein, 1998):

- Gather information
- Sort information
- Focus on the situation

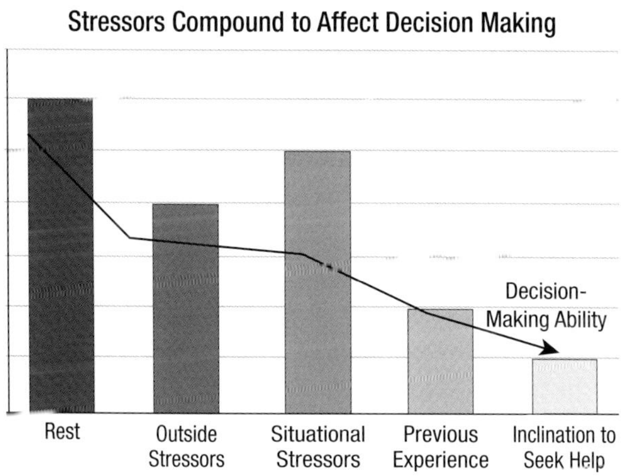


**Fig. 7–1.** How stress affects decision making. Further, how we react to the situation can increase or decrease that stress.

The span of control ICS provides becomes critical to decision-making processes when put in this context. Having too many people under one's control negatively increases stress, which leads to task overload. ICS training typically says that the ideal span of control is five people but notes that a span of three to seven may be acceptable in certain situations. Although ICS training often attributes a different span of control to a person's experience, other factors need to be taken into consideration as well.

A person completely capable of handling seven people at an emergency one day may be capable of handling only four people at a very similar incident a few months later because of outside stressors or lack of rest. Outside stressors can come from a variety of places, such as a child having trouble in school, a family member or close friend who is sick or recently died, financial problems, or marital problems. These outside stressors combined with a lack of sleep at the second incident can quickly push a person into task overload, which is also known in emergency services as analysis paralysis (Weick 1990). Therefore, the span of control built into ICS improves decision making by reducing the possibility that a person will experience task overload while it also improves accountability by reducing the number of people for whom each supervisor is responsible.

## Deferring to the expert

Given the multitude of tasks emergency services are expected to provide, it is impossible for any one person to be competent at all of them because it takes up to ten years to truly be an expert in a specific field (Gladwell 2008; Klein 1998). Also, technical knowledge changes constantly, so it becomes impossible to stay current with all of the latest, relevant information unless it is a skill or ability used frequently. One characteristic of experts is that they recognize their own limitations. Therefore, a department striving for success is willing to accept input from someone who knows more about a topic than the person in charge. This is also true on an emergency scene. A supervisor, whether it is the incident commander, a section chief, or a supervisor, is encouraged to take input from a specialist; the worse the consequences will be for a bad decision, the more beneficial it is for the supervisor to defer to an expert. This is yet another example of when information flows up the chain of command rather than down. Nevertheless, it is critical to the successful outcome of an incident because the specialist is put in a position to assist with decision making.

Hazardous materials and technical rescue calls both serve as good examples. The incident commander is not likely to be proficient at both specialties. The incident commander has the option to designate either a specialist to help by providing specific information or a knowledgeable person as the branch manager or group/division supervisor. This is a way to push decision making down the chain of command to those who are most capable of performing the task.

## AFTER-ACTION REVIEW

As was discussed previously, ongoing learning is an integral part of any organization striving for success. An important piece of ongoing learning and higher-reliability organizing is reviewing incidents. This is a formal step of ICS. A department that has developed a learning culture is going to be more successful at determining what did and did not go well, but every department has the opportunity to learn.

An after-action review can be formal or informal, but either way it should attempt to answer four questions:

- What was supposed to happen at the incident?
- What actually happened at the incident?
- What went well, and why did it go well so it can be repeated in the future?
- What did not go well, and how can it be improved in the future?

A good facilitator is critical to a successful after-action review. An effective facilitator will encourage an open environment. One tool that works well for developing an open environment is having a senior member be the first to answer the four questions. The senior member, determined by either rank or informal leadership exercised in the department, sets the tone. An effective facilitator will also ensure that everyone has the opportunity to contribute. It is often the quiet person who has something very enlightening to interject.

Time is of the essence for an after-action review. Whether organizing an informal review—often referred to as a tailboard talk—or a formal review to involve more people, it is important to conduct it as soon as possible. Time is important because:

- The only way to correct actions or behaviors that result from or contribute to RPDM is to correct them quickly, typically with a tailboard talk.

- The longer you wait, the more people begin to remember what they should have done in place of what they actually did, so the final interpretation may not be entirely accurate.
- People forget certain events that happened as they focus on the bigger picture and less on the details.

Setting the stage for a formal after-action review helps those who were not at the scene to understand the situation. Simulation is one way to add experience for RPDM, so the more information that is available to help observers place themselves in the situation being reviewed the better. Some tools that can help set the stage for an after-action review include:

- The audio recording of the radio traffic
- Aerial photographs from map websites to give an overall view of the scene
- Street-view photos from map websites to give more specific views of the scene
- Pictures or video of the scene during and after the emergency (a dash cam in the command vehicle is a great source of video for training)
- The ICS forms or any documentation that notes the assignments given

It is helpful for the facilitator to review the audio recordings and ensure that the assignments given on the radio align with the assignments noted in the ICS forms. This also helps establish a time line for the incident. After giving an overview of the emergency scene with the online mapping pictures, playing the dispatch and initial on-scene report for the after-action review helps put everyone into the right mind-set.

The single most important thing the facilitator can do is create an environment where people feel free to speak. Learning can occur only if people are willing to admit that a mistake was made, so discussions should not become personal attacks on decisions that were made on the scene. It is disingenuous to say, "I would have done this or that," after the fact because we all know what happened by that point. For example, it is counterproductive to say, "I would have never put a crew on the roof," if a crew was nearly caught in a roof collapse. Of course no one would put a crew on the roof if they knew a collapse was going to occur.

Nevertheless, it is important to discuss close calls. It is important to share the signs that alluded to changing conditions that may not have been noticed until after the fact. Framing the discussion in terms of how it could be done the next time to keep the same thing from happening, helps the conversation remain productive.

Last, it is important to discuss what went well. Too often, the focus is on what did not go well. However, it is important to reinforce the positive actions taken so that those actions continue and are shared with others who may not have had the same experience.

## INCIDENT COMMAND IS A CULTURAL BEHAVIOR

Bigley and Roberts (2001) have researched the idea of using ICS in private business. They looked at a Southern California department that had used ICS for some time. They found that incident command is as much a cultural behavior as it is a system. ICS encapsulates the thought process that becomes engrained in firefighters and then provides tools for sharing that thought process.

ICS is so flexible that it is helpful to use not only on incidents but also when organizing other operations. The thought process and organization provides the information people need to achieve daily tasks or nonemergency events because it provides objectives and information that help people in the department coordinate their efforts.

Just as we have muscle memory for hands-on skills, we also have brain muscle memory for decision making. By developing a level of understanding to unconscious competence, we are developing our brain's "muscle memory." By doing this, members using the ICS know how each of their roles fits into the overall picture, when to communicate relevant information, and how to adjust with changing conditions without much conscious thought. This thought process becomes so automatic that people act on the thought processes without even thinking about it. Many who use ICS regularly have experienced a similar situation with their spouse. When a problem arises at home, a person with unconscious competence sizes up the situation, notes a desirable outcome (objectives), and begins solving the problem immediately (strategies, tactics, and resource assignment). However, the spouse is still in the exploratory stage of figuring out the problem. This is where the friction typically arises because one person is still trying to comprehend the problem while the other is moving forward with a possible solution understanding that adjustments may need to be made as more is learned. The friction is not actually a result of the spouse trying to solve the problem but the failure to communicate the thought process and gain a shared understanding.

### Case in Point

You and your spouse are leaving home to go to dinner. It is Friday and there is a lot of traffic, making you late for a reservation. You come up with plan A to avoid traffic and begin executing it without any communication. You know plan A may not be perfect, but you also know there are options for adjusting as you go. Your spouse is developing plan B but is not as familiar with the area. Your spouse is trying to comprehend your actions while still developing plan B with what seems like more unknowns because your spouse sees fewer options for adjusting. It leads to a disagreement over the route you take to dinner, and the pressure of being late for your reservation increases the friction. The problem here is not either plan. The problem is that you did not run plan A past your spouse prior to executing it and explaining that there are alternative options if it does not work out.

This is a simple example that illustrates the situation described above in terms of something we have all probably encountered before.

Developing the entire department to this level is difficult because it takes time. The normal flow of personnel from retirements and hiring prevents this. However, ICS is designed to accommodate this. The new firefighter is the one who should be task focused; the captain, who should have unconscious competence, has a broader focus of what is happening nearby and what should be happening in the near future; and, the incident commander is considering the whole picture. Combining this level of operation at every incident with training and realistic scenarios develops a cultural behavior that allows ICS to work with greater reliability. The more engrained the behavior is, the more reliable ICS becomes, and the more successful an organization will be at accomplishing its goals.

## REFERENCES

Bigley, G. A., K. H. Roberts. 2001. "The Incident Command System: High-Reliability Organizing for Complex And Volatile Task Environments." *Academy of Management Journal* 44 (6): 1281–1299.

Deal, T. 2010. *Beyond Initial Response: Using the National Incident Management System's Incident Command System*. Bloomington, IN: AuthorHouse.

Gladwell, M. 2008. *Outliers: The Story of Success*. New York: Little, Brown and Co.

International Association of Fire Chiefs. 2010. *Fire Officer: Principles and Practice*. 2nd ed. Sudbury, MA: Jones and Bartlett.

International Association of Fire Chiefs and National Fire Protection Association. 2009. *Fundamentals of Fire Fighter Skills*. 2nd ed. Sudbury, MA: Jones and Bartlett.

Klein, G. A. 1998. *Sources of Power: How People Make Decisions*. Cambridge, MA: MIT Press.

Stowell, F. M., and L. Murnane. 2013. *Essentials of Fire Fighting and Fire Department Operations*. Upper Saddle River, NJ: Brady.

Weick, K. E. 1990. "The Vulnerable System: An Analysis of the Tenerife Air Disaster." *Journal of Management* 16 (3): 571–593.

# 8

# YOU KNOW TO LOOK OUTSIDE THE FIRE SERVICE FOR IDEAS—BUT WHERE?

## CHAPTER OBJECTIVES

- Introduce crew resource management (CRM) and its concepts
- Provide tools to begin using CRM
- Enhance understanding of how CRM has a positive influence on human behavior at critical operations
- Describe how key CRM components support higher-reliability organizations

## IDEAS FROM OUTSIDE THE FIRE SERVICE

Information is invaluable—and is often the source of groundbreaking ideas. It is the key ingredient to making good decisions and to achieving success. It is a part of everything we do every day. This is the reason many high-profile leaders are avid readers. However, most of these leaders do not work in an emergency environment that is often loud, dark, overwhelming, and time-compressed. By contrast, in the fire service, we usually do not have the luxury of time to fully research critical decisions that affect life safety of both our own people and the people we serve. The bottom line is that good decisions based on accurate information lead to success—as a profession, as a fire department, as a crew, and as individuals.

### Discussion Questions 8–1

1. Where do you get information? Where should you get information?
2. Are the ways you get information accurate and current? How do you know?
3. What are some industries with similar characteristics, products, services, or operations?

But what is information? Information comprises data, knowledge, communication, ideas, research, theories, techniques, methods, and even experience. Where do we get it? We get information from books, training, school, professional publications, the media, online, research, experimentation, and again through experience. There is no shortage of access to information these days, but are we looking in the right places? Did you list every one of your crew members as a resource for information? Were they your first thought, your last thought, or did you not even think of them?

As humans and creatures of habit, it is tempting to fall into a routine when looking for information—we may even wait for it to find us. We love comfortable, but comfortable does not get it done. The most successful people, crews, fire departments, and professions aggressively seek out information anywhere and everywhere. They treat information like it is worth its weight in gold—because it is. When we look only in the easy places, that often means we are looking within our own departments. In some departments, there are not enough active learners who venture outside their own department to gather information. Consequently, their departments may enjoy moderate levels of success but will never be cutting edge or leading because they are taking in only what exists in their own fishbowl realm. This has been called "inbreeding" by some.

Some departments take it a step further by using the tried and true method of copying someone else. There is nothing wrong with this, and although it works well, our hope is for our readers to become the ones who are copied. There are even fewer brave souls who risk being the first bloodied in battle as they try to bring about that evil word *change* with new ideas, techniques, or systems from outside our industry. They may not always be successful; in fact, they will fail more often than not. Still, eventually they will succeed because there are lessons for improvement and useful ideas within the failure. When they do take these lessons and ideas for future attempts at change, watch closely because you are experiencing greatness—and experiencing true leadership. These are true leaders: the people who are successful for themselves, their crews, their departments, and their professions.

Not all of our new ideas, trends, and information will come from inside our industry. We must also observe other industries, especially those whose products, operations, or services resemble those of our own industry. What characteristics does the fire service have as an industry? We provide a wide range of services, so we are primarily service oriented. We work as crews. We use a paramilitary hierarchal structure. We provide many of our services in hazardous environments where mistakes and poor decisions can have a high life safety price (high risk and high consequence). Our industry can have hours or days of little action that are suddenly interrupted by incidents of great magnitude. Our industry focuses intensely on public safety education, so we are also teaching oriented. Depending on what fire department you work for, there may also be many other characteristics involved with the many other services we provide such as EMS, technical rescue, hazardous materials, and wildland urban interface. So, what are some other similar professions?

## Exercise 8–1

| Profession | List similar characteristic(s) to the fire service |
|---|---|
| Law enforcement | |
| Military | |
| Airlines | |
| Education/teaching | |
| Emergency medicine | |
| Air medicine | |
| Service-oriented businesses | |
| Others? | |

The point of the previous exercise is to get you thinking about other industries from which we can get information. Some of the best sources of information are professions with similarities to ours, and there are numerous alternatives. Being open minded by studying industries with similar operations can greatly benefit our own operations.

Although the airline industry may not seem to have much in common with the fire service at first, it is a service industry with important similarities from which we can learn. It is important to consider the traits and employee behaviors instead of the actual tasks being performed. Airline personnel work in a high-pressure environment focused on making time-critical life safety decisions. They are allowed very little margin for error because the potential mistakes would be so costly. They check, operate, and maintain high-performance equipment. They encounter hours of boredom separated by moments of sheer terror. They work as crews. They work in a paramilitary hierarchical structure with designated ranks. In some cases, they even attract the same types of people (particularly those who are attracted to some degree of excitement, action, and danger). Does this sound anything like the fire service? It does to us, and that is why Crew Resource Management (CRM) is central to this chapter. CRM is a system the airline industries adopted to reduce accidents and has since been adopted by many other industries with similar traits.

Although we have many similarities, one big difference in emergency services is the lack of an information-gathering system used to improve operations and reduce accidents. The airlines have focused intensely on learning (gathering information) from past mistakes since the early 1980s. The airlines have had a subtle advantage over us in that they were forced to make changes, because the large number of accidents they were encountering were making people question the safety of flying. If people were to stop flying, the airlines would go out of business; consequently, the airlines had no choice but to address their safety problems. The airlines took a close look at their accidents and found that airplanes were not crashing because of mechanical problems; instead, airplanes were crashing because of people problems. In other words, airplanes that crashed were operationally sound, and it was the people inside them who were responsible for the accidents.

CRM was developed as a solution to deal with the human side of operations. At first, airline pilots rejected and disparaged this change as "touchy-feely training." After years of working with CRM, the pilots' resistance has finally waned. This represents a normal reaction to change. Over time, CRM has become a part of normal operations. New pilots do not know operations without CRM; because it is normal for them, they do not fear, reject, or fight it. Since the initial introduction of CRM, the airlines have gradually reduced the number of accidents—by over 80%.

This change did not occur overnight; it has taken the airlines years to get to this point. Similarly, it would also take the fire service years to realize such results; nevertheless, if the fire service could eventually reduce our accidents by 80%, it would be worthwhile. Although well intended, statements such as "be safe," "maintain situational awareness," and "make good decisions" are useless without explanation, true understanding, and function; we need to connect the dots to find the underlying meaning of these broad statements and implement it. CRM explains what these statements actually mean and how they can be achieved (International Association of Fire Chiefs 2003).

## Discussion Questions 8–2

1. What do the statements "stay safe," "maintain situational awareness," and "make good decisions" mean to you?
2. How do you accomplish their intent?

---

We teach people to make good decisions by teaching them the how *and* the why. Simply making a list of safe actions is well intended, quick, and easy; however, it is not effective. Lights, buzzers, SOGs as voluminous as an encyclopedia set, and punishment—although important in each of their own ways—oversimplify complex human interaction in critical situations occurring in unpredictable environments. Quick fixes only cover up a problem, letting us think we took care of it. In the meantime, the problem continues to grow because we only swept it under the rug and invariably will resurface at a later date—possibly with even more severe consequences. There are no true solutions that are quick and easy: Real solutions require continuous work and focus; real solutions are ongoing; real solutions require constant training and learning.

To focus and continually improve, industries develop systems. Although the word "systems" may not be appealing, please bear with us. Systems are how HROs improve, and many of their systems have been proven to work. Although none of these systems are simple and they all require commitment to indefinite improvement, the rewards are worth it.

CRM is a system that focuses on the use of all available resources. CRM is a tool to optimize human performance and reduce human error through the use of all available resources. It works to make the entire crew better, not just the leader. It has been proven to work in several industries, and there can be a strong correlation drawn between how long they have been using it and the success levels they have had with it. Basically, the longer an industry has been using it, the fewer accidents the industry experiences. The following seven segments cover the major parts of CRM, as well as how it works and how to implement it. Some of these topics have been covered in greater detail earlier in this book, so here they are reviewed in brief or from a different angle to spur thought and discussion as it is relevant to CRM.

## WHAT COULD POSSIBLY GO WRONG?

Decision making is part of almost everything we do and can produce both good and bad outcomes. Because decision making is involved in so much of what we do, it has a great impact on emergency operations and the actions we take when mitigating emergencies. Thus, it behooves us to make the best decisions possible, and as we stressed earlier, this requires study and practice. To improve decision making, we must first learn how to make decisions and what affects the process. How we make decisions depends on how we interpret the world around us.

### Putting together the puzzle

Everyone is different because of their different life experiences—namely, different upbringings, hobbies, interests, educations, friends, teachers, and work experiences. This is why each of us sees and interprets the world around us differently. Each of us has a unique picture that we see all day, every day, of the world around us. This is important for emergency responders to understand because it means that at every fire, automobile accident, technical rescue, hazardous materials scene, and EMS incident, each person is seeing a different picture in their mind. Most likely these pictures are similar, but they differ in the details, which affects the way the picture will be interpreted—and the possible solutions to the problem. In other words, we cannot see everything at a scene. We overcome this by realizing that it happens so that we can place effort on getting more of the picture.

Imagine a jigsaw puzzle depicting a fire (fig. 8–1). Each one of us has one or two of the pieces, hence only part of the picture. When we put all of our puzzle pieces together, we see the whole picture. Although this is a simplified example of everyone's unique perception, it illustrates how important it is to collect and share information. It shows that when we all share information—the picture in our minds and how we interpret it—we get a much more complete picture of everything that is actually happening around us at all of the emergencies to which we respond.

CRM is a tool that is designed to do this. CRM teaches us when to seek out the expertise of those around us. CRM also teaches us when and how to speak up because we have information relevant to the situation.

## Veteran versus novice

Recognizing that individuals see a similar picture but different details at emergency incidents, next we will look at how experience in fire and emergency response affects what we see and how we interpret it. For simplicity, we will break experience into two categories: novice and veteran. The more experience a person gains, the more they move toward the veteran level. Even though this implies a discrete division between these experience levels, in reality, there is no such dividing line for experience levels; still, there are several rough measures to help find where you fall in terms of decision making.

Several measures may differentiate a veteran from a novice. The rough guideline for possessing veteran interpretation is responding to 2,000 emergency incidents and being the incident commander 200 times (LeSage, Dyar, and Evans 2011). Depending on the size of your department, your run volume, and your role, the time it takes to reach this will vary. In the airlines, the veteran level is often around 1,200–1,500 total hours of flight time, with at least 200 of the total hours as the pilot in command. For pilots, this time represents engine running time only, so they have a lot of other time in preflight planning and preparation that its built into the hourly requirements. This is the base

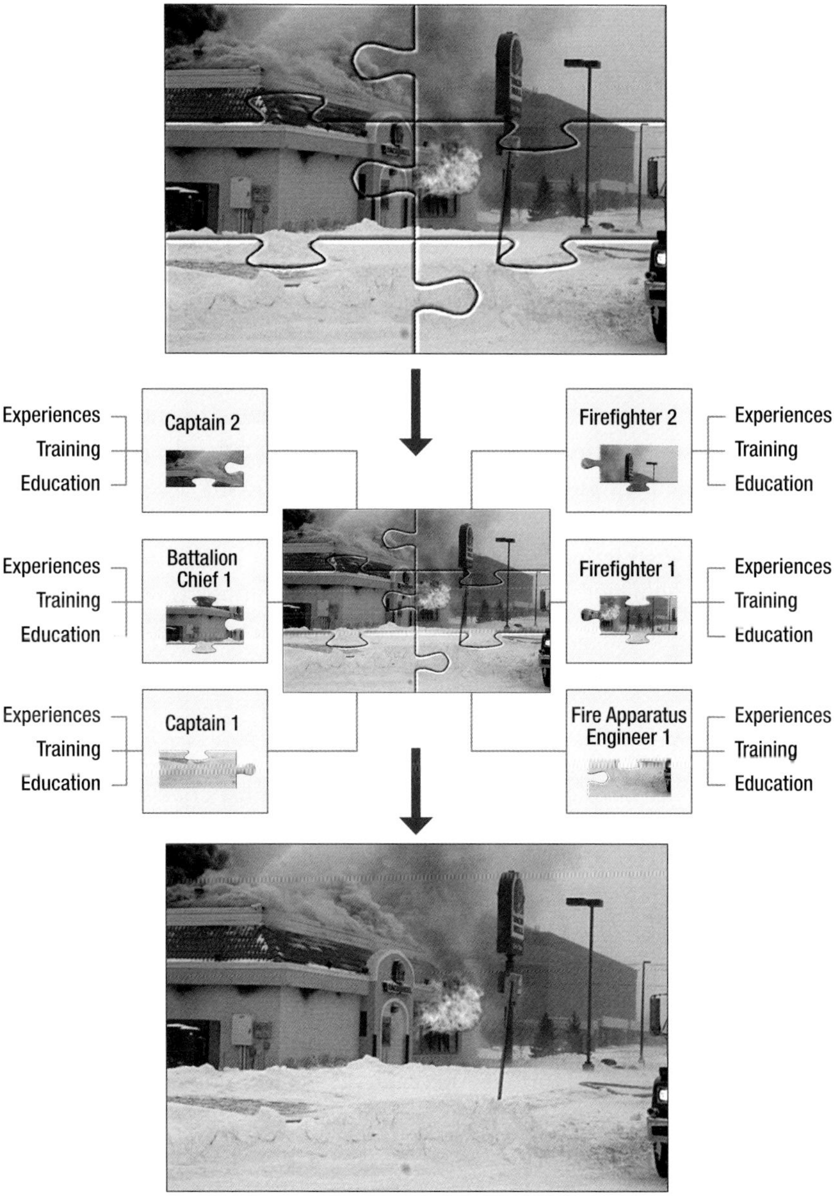


**Fig. 8–1.** Illustration of the different pieces of an emergency scene people might see, which when put together gives a more complete picture of what is actually going on

level of experience for getting hired at many major airlines. Yet another guideline says that an expert is someone who has practiced a trade for 10,000 hours (Gladwell 2008). The point is that your experience level affects how you make decisions. Now that we have defined experience, let's look at how each level makes decisions.

Experienced, or veteran, firefighters have an ability to process and prioritize large volumes of emergency incident information very quickly. This skill helps veterans to not only quickly mitigate emergencies but also quickly recover from mistakes or unexpected occurrences.

As discussed earlier, veteran decisions are based on RPDM, whereby the brain scans through past successful experiences (memories), finds a match similar to the current event, and then uses the decisions made at that incident to guide current decisions. RPDM is a very effective thought process (Klein 1998) that allows veterans to more easily ignore or minimize distractions. However, veterans using RPDM tend to miss the smaller details of an incident because they overlook those details in the process of finding a matching incident from before because of their focus on the bigger picture.

Newer, or novice, firefighters do not yet possess the ability to process large volumes of emergency incident information quickly because they do not yet have enough emergency incident experiences for their brains to reference using RPDM. This does not mean that novices cannot make good decisions or that they do not have anything important to share. Novices still see a picture and must be able to react to it. If they do not use RPDM, then what do they use to make their decisions? Consider the references to which novices would have been exposed early in their fire service life. Textbooks, SOGs, and training are what novices use to make decisions. This equates to their processing information in a very technical format and results in novices encountering three times more decision points than veterans do (LeSage, Dyar, and Evans 2001), slowing them down. Novices are also more easily affected by distractions, leading to even more decision points. This is why it takes longer for novices to recover and why novices may experience analysis paralysis once in a while. Freezing in place is the outcome of their brain's attempt to make sense of the information and decide what to do with it. For example, where do you start when everything in sight is on fire and you have never been to a fire before? Novices see more detail because of their technical view of emergency incidents, and these details should not be overlooked or undervalued.

It is crucial that the detailed viewpoint of novices be shared with everyone at an incident because a novice may be the first one to see spalling, sagging, or puffing smoke at a fire. However, novices have to be taught to share this important information; otherwise they might assume that all of the more experienced veteran firefighters have seen it too but discounted it as something not to be concerned about.

Veterans must also learn to seek out this detailed viewpoint from novices to help get a more complete picture of the incident. By combining the big picture seen by veterans with the detailed picture seen by novices, those running an emergency incident will get the whole picture that includes all the information—or at least the viewpoints of all those on scene. Now that you have the whole picture, or at least more of it, you can make better decisions.

## MEMORY AIDS

The human mind is not perfect. It is susceptible to forgetting, misremembering, misinterpreting, and more. This does not help when time-critical life safety decisions need to be made on a daily basis. To make better decisions and combat these common fallibilities of the human mind, many in the fire service use memory aids. These include acronyms, models, and guidelines for operations. One of the most common is "RECEO-VS" (rescues, exposures, confinement, extinguishment, overhaul, ventilation, salvage). This acronym was designed to help the fire service prioritize actions at a fire. Newer adaptations of RECEO-VS reflect advances in our understanding of fire behavior. Another aid comes from ICS and ranks the priorities: life safety, incident stabilization, property conservation, and environmental concerns. There are also decision-making models such as risk versus benefit (i.e., risk a lot to save a lot, and risk a little to save a little). In this model, we are taught to analyze the level of risk and compare it to the level of reward. SOGs assist with decision making by providing initial guidance for specific situations including high-risk incidents such as high-rise fires.

These types of memory aids work well for those who are trying to move into the veteran decision-making role—but they are only tools. To be useful and fully understood, the tools must be used in practice. The more you practice with these tools, the more effective you will become as your understanding of their meaning increases. Memory aids can also be detrimental, as shown in the Cerro Grande example (see chap. 6), when their meaning is not fully understood.

## STRESS

Stress surrounds us every day. It comes from a never-ending list of sources. In any discussion involving decision making, stress must also be included because it directly affects our ability to make decisions. As emergency responders, we usually do not have the luxury of time to make our decisions. A lack of time means that we probably cannot fully research the best direction. This increases the impact that stress has on our ability to make decisions well. This is because there is an inverse relationship between our ability to make good decisions and our level of stress. As stress levels increase, our ability to make good decisions decreases. The higher the stress level is, the more significant the impact will be on your ability to make good decisions.

This affects every human. Even the most experienced, smart, and well-educated leaders in the fire service have breaking points beyond which they become overwhelmed and ineffective because they are still human. They may be able to handle more stress than most, but every human's ability to make good decisions decreases as stress increases. When stress levels get high enough, humans lose cognitive functions, physical functions, and auditory capabilities. Consider the stressors that emergency responders face, which go beyond normal daily work and family. What happens to your pulse when the tones go off? How about when you get in a big red truck with flashing lights and loud sirens? When you arrive at a scene where fire has self-vented from a second-story window? When a mother runs up to you screaming that her child is still inside? The point of these examples is that emergency responders encounter very stressful situations, and that is part of the job. To make better decisions, we must try to reduce our stress. Although causes of and ways to reduce stress have been covered in detail earlier (see chaps. 1 and 7), here is a summary of these methods:

- Use an aide
- Move to a more comfortable work environment—out of the weather and noise (e.g., inside a vehicle)
- Write information down
- Delegate
- Maintain sufficient resources
- Engage in preplanning
- Train—particularly scenario-based programs
- Encourage more communication to build a better picture
- Live healthy—rest, eat, exercise
- Control or slow down your breathing

For us to reduce stress, we must recognize that it is affecting us. To do this, we create a personal list of flags. Flags are like a lightbulb that goes on in your head alerting you to a potentially hazardous situation. They come from previous experiences and by studying where and when you are likely to encounter problems in your operations. An example involves the overall most stressful type of incident for firefighters—namely, incidents involving children. Knowing this, we have made incidents involving children a flag because we know we will be more likely to speed to get there, more likely to take great risks once we get there, and less likely to make solid decisions. Another flag might be when you hear a lot of ums and ahs over the radio, because such verbal pauses could signify that the person making the transmission is getting overwhelmed.

## Exercise 8–2

Make a list of flags for yourself to alert you to increasing extreme stress levels or an incident about to go bad:

- 
- 
- 

- ¤
- ¤
- ¤
- ¤
- ¤

This should be considered a partial list of flags to add to continually.

---

## WHAT THE HELL IS GOING ON?

Now that you are equipped to make better decisions, or at least to recognize when you are likely to make less than good decisions, we can try and figure out what the hell is going on around us. Emergency incidents are chaotic scenes where it can be very difficult to obtain, let alone maintain, situational awareness. Has anyone ever told you, "Maintain your situational awareness"? What does that mean? How do you do it? Hopefully, after reading this section, you will not only be able to explain what it means but also have learned how to gain it and, perhaps more important, recognize when you are losing it. Situational awareness is when what you think is happening and expect to happen in the near future is what is really happening—that is, your perception (the picture in your mind) matches reality. If something does not make sense, then you do not have complete situational awareness.

Situational awareness is gained by gathering information primarily through sight and sound and then interpreting that information as it relates to the world around us. To gain situational awareness, we must first gather information. At emergency incidents, this is done by gathering preplans, dispatch information, bystander information, and crew information. A good cross-check on your situational awareness is to ask another emergency responder what their interpretation of the incident is. If your and their interpretations match closely, then that is a good sign your situational awareness is solid. This process is constant. One way to do this consistently is to compare your interpretation of the incident with the incident safety officer's interpretation after the safety officer has completed a walk-around to see how closely they match. If you stop gathering information about an incident, then you lose your situational awareness.

## Exercise 8–3

List some ways to gather information to gain situational awareness:

- ¤
- ¤
- ¤
- ¤
- ¤
- ¤
- ¤
- ¤
- ¤

---

The fire service seems to do well with initially gaining situational awareness at emergency scenes. That is what we are taught to do from before the alarm bell rings, through the en route phase, to the arrival on scene. Where we often have trouble is recognizing when we are losing situational awareness in the middle of an incident. If losing situational awareness is not caught quickly, it will often be too late to undo previous events by the time it is apparent to you and the rest of the world that what you thought was going on really was not.

This relates closely to the previous section on decision making because stress and information processing play key roles in situational awareness. Recognizing situational awareness may seem easy, but it is complex and difficult. Like anything else, this recognition is improved with practice and experience.

As previously discussed, it may be easiest to start by identifying flags that indicate when you are losing situational awareness or by asking for

CAN reports to see if the reports match what you are seeing. Just as we used flags to help us identify potentially hazardous situations, we can also use flags to recognize when we are losing situational awareness. The following list comprises situational awareness flags that we have identified:

- Lack of communication (it is too quiet)
- Lack of information (no CAN reports, no preplanning information)
- Something about the incident does not make sense
- Unfamiliar with operations (have not seen something like this before)
- Did not establish and name command; lack of incident organization
- Do not know where every responder is (lack of accountability)
- So many distractions that you cannot think straight
- Did not complete a walk around

The preceding list is not complete and should be added to as you think about, practice, and discuss situational awareness with others on your crew and in your department.

Solutions for maintaining situational awareness address the flags identified. By determining where we lose situational awareness, we can find the solutions for gaining and maintaining it. Solutions for maintaining situational awareness include:

- Build your own list of flags
- Establish command and complete a walk around every time
- Constantly communicate and gather information
- Use an aide or scribe
- Reduce distractions (reporters, positive pressure fan noise, etc.)
- Reduce stress (eat better, get sufficient rest, get out of the weather, etc.)
- Prepare appropriately for weather or get out of weather (or both)

- Use other people to gather information
- Train using scenarios with distractions
- Review incidents
- Use near-miss reporting

As we continue, you will find common threads between decision making, situational awareness, and communication. Some points are repeated, but these bear repetition because they are crucial to successful operations in high-risk environments such as the fire service. Stress is one of these repeated concepts because it affects decision making, situational awareness, and communication. Use this as an opportunity to fully explore and understand how great an impact it has and the havoc it can wreak at our emergency incidents because it is intertwined with just about everything we do. Even while we cannot avoid stress, by gaining an understanding of how and when it affects us, we can learn to reduce its effects to consistently attain more successful outcomes.

If this seems impossible, remember the airline example. Over 30 years, they have reduced their accidents by 80% by looking at human solutions, and human solutions begin with understanding human factors. As a crew's stress level increases, their cognitive and physical abilities will be lost (e.g., the ability to process auditory signals) and they will miss important information, which will cause them to lose situational awareness and make poor decisions. If incident commanders are untrained in human factors, then they too will lose the ability to process information and situational awareness in these circumstances but may not realize it. Thus, when a Mayday is called, the incident commander may not hear it, may not communicate with the transmitter, or may not address it in as timely a manner as possible. What you do not know can and will kill you or someone else. This is why CRM and human factors are important in the fire service.

The next section is on communication, which is the third critical piece of operations—completing the human operations triangle of decision making, situational awareness, and communication. These three pieces all directly affect each other and are themselves affected by individual human factors of the people working together at emergency incidents.

# WHAT?

Every time I call my son on the phone, he answers, "Hello," and I respond with, "Hi, how are you?" To this, he replies every time, "What?" Sometimes he throws in a, "Wait, what?" I am pretty sure it is because he is not paying attention, and his mind is distracted by a computer, television, tablet, or something else. I often wonder how much of our conversation he really hears when he is already asking, "What?" after three seconds.

Communication plays a key role in emergency response. When we ask those in the fire service if communication has ever been an issue, they reply that it is an issue at almost every fire incident and especially the large fire incidents. Yet, how much time do we spend learning about communication and then training on it? Surely we must give it a high priority in training, because it is one of the most common issues we have, right? Even though communication can play and has over the years played a major part in firefighter line-of-duty deaths, we tend to give it little attention. After reading this section, you should understand not only the true importance of communication but also some of the barriers to effective communication and how to practice effective communication by using the PACE (probe, alert, challenge, emergency intervention) process.

## Too much or not enough communication?

We would argue that the fire service as a whole, through our culture, has taught us that too much communication is bad. In some ways, we are even taught to stay off the radio, to keep from clogging up the channel with chatter in case there is a Mayday. To the contrary, though, we strongly believe that the fire service does not communicate enough and that continual communication over the radio represents one of the primary sources of information for us at an emergency incident. Radio communication is one of the best methods for quickly relaying time-sensitive information both up and down the chain of command. In previous sections, we have shown that information helps us to make good decisions and increases our situational awareness. If firefighters are unable to transmit information, then decision making is handicapped.

An example of success from the airline industry illustrates the importance not of only good communication but of communicating a lot. If we think of communication as transferring information, then transferring information as much as possible during emergencies makes for the best possible result.

## Case in Point

### *United Airlines Flight 232, 1989, Sioux City, IA, DC10*

An example used earlier taken a step further shows how good communication leads to improved situational awareness, which leads to better decisions, which leads to better outcomes. A United Airlines DC10 experienced a total hydraulic failure while in flight. The aircraft declared an emergency and crash-landed in a survivable manner at the Sioux City, IA, airport. The crew worked together to do what many in aviation would consider impossible. The average number of communications over the entire incident was 31 per minute, and at the height of the incident, there were 60 communications per minute. The captain and crew have been celebrated as an example of CRM working even better than could be imagined, because in attempts to duplicate their achievement in flight simulators, no one has ever landed successfully.

Good communication (information) = Situational awareness

Situational awareness = Good decisions

Good decisions = Better (safer) outcomes

Our hypothesis = More communication is better

When the airlines started their research into why airplanes crash, they found one of the leading reasons was the lack of assertiveness in junior officers to speak up or take action (or both). They found that the reasons for this were rooted in the hierarchal structure of the airline crew. At the time, the airline flight deck consisted of a captain, a first officer, and usually a second officer. The captain of the aircraft was treated—and often acted—like God. Captains were not to be questioned or interrupted, let alone corrected. Over the years this culture grew and strengthened; new pilots were trained into this culture, and this culture was reinforced over time on the flight line. When the airlines, together with the National Aeronautics and Space Administration, finally started looking deeper into the causes of their accidents they found that this culture was a major factor; it had become so ingrained that first and second officers failed to speak up even when they knew something was wrong, even when this was something that could lead to their own deaths. Over the past 30 years, the airlines have dramatically changed to a culture where new people not only are taught that they are expected and encouraged to speak up but are actually told that they are paid to speak up. This teaches new pilots from day one how important they are to the safety and success of operations and lets them know that even though they are new and have a lot to learn, they are smart, valuable, and capable employees.

Unfortunately, in contrast, the fire service still teaches that the only people on the fire ground who can stop an operation are the incident commander and the safety officer. These two positions have been elevated to a status similar to that of the old airline captains. Partly this has happened because the fire service tends to focus on the technical side of communications: addition of more tactical channels, repeaters for high rises, and so forth.

We must address the real reason why communications do not happen when they should. In a fire, the people inside are often the first to become aware of a safety issue; yet, we currently discourage them from trying to stop an operation if they are not the safety officer or the incident commander. We propose that everyone at the scene be taught to act as a safety officer; that is, if you see something that could get someone hurt or killed, it is not only a good idea but also your duty to speak up. (You should still always have a designated safety officer.)

## Barriers to communication

The fire service hierarchy itself is a common barrier to communication (Carley 2013). However, there are many other barriers, including anything that gets in the way, reduces effectiveness, interrupts, prevents, confuses, slows, or misdirects a message (information) between the sender and the receiver. In emergency response, even though we cannot eliminate all of these barriers, we can learn to identify them and reduce their prevalence and impact. By listing common barriers to communication, we begin to realize how many there are and learn what others exist that we never realized:

- Failure of new and junior people to speak up
- Weather
- Equipment (radios, antennas)
- Physiologic abilities (heart rate, stress)
- Language barriers
- Vague terminology
- Complacency or laziness
- Fatigue

## The PACE process

We have now familiarized ourselves with common communication issues and learned how important it is to proactively teach new and junior people to speak up. One method that has been developed to build better communication skills is the PACE process (Besco, n.d.):

- Probe
- Alert
- Challenge
- Emergency intervention

This process should be used by crews whenever they see something that should be communicated. It begins politely with a probing question, progresses to an alert statement, then becomes a more abrasive challenge exclamation, and finally escalates to an emergency intervention.

For those trained to use it, recognition almost always occurs at the probe level, where the conversation is most polite. A person can jump to any level that most appropriately addresses the situation at anytime. It is not necessary to start at the probe level and go through all of the levels in order if there is an immediate danger. The determining factors for which level to use are the amount of time to react and the severity of the possible consequences of not taking action. For example, if a roof is about to collapse and a firefighter sees it before an officer, it would be expected that the firefighter should jump straight to the emergency intervention by grabbing the officer by the coat and pulling him or her out.

A simple way to picture each step and how it is spoken is shown below:

*Probe?*

*Alert.*

*Challenge!*

*Emergency intervention**

Notice that the probe step has a question mark because it represents a situation in which there is time to express and discuss a concern. The alert step has a period because it represents a situation where there is a little less time for politeness, so instead of questioning whether there is a problem, you are stating that there is one. In the third step, an exclamation point implies that time is of the essence and represents yelling or shouting at someone because there is no time to be polite. The last and fourth step, emergency intervention, has an asterisk because it is a wildcard and anything goes at this point—that is, if immediate action is not taken, there will be dire consequences.

All of the statements in the PACE process are assertive statements. As such, they are practiced and spoken as respectfully as possible but in a certain format. Each PACE statement includes all of the following elements (with exception of the last, for the challenge and emergency intervention steps, because there is little or no time):

- Identifier (name or rank)
- Personal concern
- Define the problem
- Provide a solution
- Seek a response (although there may not be time for response in the challenge or emergency intervention steps, compliance should be considered equivalent to a response)

## Case in Point

Examples of each step in the PACE process are provided here.

### *Probe?*

- A respectful question or low-level statement to increase shared understanding
- Not time compressed

*Engineer Stevens, I'm concerned that you are driving too fast. I think you need to slow down. Do you agree?*

### *Alert.*

- A statement bringing a pending problem forward
- More time compressed, or the probe did not work
- Can be the first interaction

*Captain Jim, I'm concerned that you are not wearing your SCBA. I think you should put it on. Reply if you understand.*

### *Challenge!*

- A direct statement escalated to a respectful but attention-grabbing format
- Time compressed, or the alert did not work

*Stop, Gary! You don't have your flash hood on. Put it on now!*

### *Emergency intervention**

- An action taken to stop the event
- Not limited to incident commander or safety officer
- No time, or the challenge did not work

*As the firefighter physically grabs the tool of another firefighter winding up to vent a window, she says, "Stop, do not vent that window!"*

## Exercise 8–4

Develop a statement or action for each step of the PACE process for the following scenario; be sure to include all of the elements for each statement as necessary:

- Crew: Captain Stevens, two senior firefighters, and you (a new firefighter).
- Three-story, wood frame, multifamily residential occupancy (type V).
- The crew is operating on the third floor attacking the fire.
- Smoke and flames are showing from the roof.
- Fire has a long burn time in the attic space.
- A truck company operating outside reports multiple times that the roof appears to be sagging in the area of the operating units.
- The incident commander advises the division supervisor that it does not look like the crews are making headway from the outside and asks if it is prudent to continue operating.
- The division supervisor is at a decision point to stay or leave.

**Fig. 8–2.** Decision point using the PACE process

Formulate and appropriately punctuate the steps of the PACE process for the example in fig. 8–2:

*Probe:*

*Alert:*

*Challenge:*

*Emergency intervention:*

---

## Recognizing and responding to conflict

Communication at some point will entail conflict. It may even come from the PACE process used above. It is important that you be able to recognize conflict in others you work with, as well as in yourself. Here are some things to look for to recognize conflict:

- Watch body language—yours and theirs:
  - Tense
  - Hands in a fist or tightening up
  - Stiff
  - Red faced
  - Getting in your face
- Listen for verbal cues:
  - Raised voice
  - Swearing
  - Unusually quiet
- Observe an escalating situation
  - Rising pulses
  - Rising blood pressures
  - Two people squaring up to each other

When you recognize conflict, you should:

- Discuss options, to increase adaptability
- Count to ten before responding
- Base responses on the issue instead of personal feeling; do not take the issue personally
- Base decisions on data instead of emotion

Although this discussion of communication is necessarily brief, it has nevertheless served the purpose of detailing the most important aspects of communication that affect emergency responders. Much as communication affects situational awareness, it also affects organizational culture. As previously mentioned, the airlines found that their communication problems were closely tied to their organizational culture; for example, it takes people with a can-do personality who like excitement and can control their fear to pilot airplanes. Similar personality types are drawn to certain careers, including the emergency services. Pilots are no exception, and when their personalities were combined with a hierarchal structure, it manifested in a culture where the captain was always right and no one else had anything to contribute.

Such an organizational culture is ripe for hazing and intimidation. It has led to many accidents in both aviation and the emergency services because hazing and intimidation affect communication. This could continue as long as everyone holds the attitude, "I'm doing this to you because it was done to me." This is not exactly logical when looking in from outside, but then many things we have done in the past have not been very logical. Times have changed, though, and there is no room for hazing and intimidation in the modern business world. The airline industry was fortunate in the long run to have been forced to overcome communication barriers, because otherwise the public would not fly anymore and that would have hurt the airline's business. The fire service has not been forced to deal with this same situation because our accidents generally do not affect the general public. Therefore, we as a fire service are going to have to choose to make these changes.

# WHY?

Organizational culture represents who your department is, how the people in your department act, what your department does, and what your department does not do. Culture encompasses our traditions and the stories that shape our behavior. Many of the things we do today came about because of stories about someone who did something similar in the past. These stories affect our decision making without our even realizing it.

If you look deeper into what we do and why we do it, you will often find that there is no longer a good reason for it. On the one hand, some of what we do now that we did years ago is still important: We still fight fires; we still honor those who have gone before us. On the other hand, much of what we still do is not important anymore and is sometimes flat out wrong. One example of the latter case involves a fire department that required that its crews go to the apparatus floor at 2100 hours sharp every night to wash the tires and pick the rocks out of the tire treads. This went on for years. Through the years, a few brave souls asked why but were reflexively told to be quiet or that it was "because we have always done this." Eventually someone asked the chief why the tires were washed. The chief was smart enough to think about a useful reply and discovered that they had washed off the excrement and removed the rocks from the horses' hooves every night in the distant past. Thus, the practice continued long after the horses were gone, even though it obviously did not fit and was outdated! So how do we ensure that our current operations and practices are still current and important? We do this by teaching everyone to ask why. Remember that the single most important behavior for an HRO to develop in its people is the drive to ask why. Everyone in your department needs to be encouraged to ask why and how to do it genuinely and politely. However, simply asking why works only if everyone is also taught that their reply must provide a good reason. Good reasons include because it is safer, cheaper, more effective, or quicker—or even "I don't know why, but let's find out." Good reasons do not include, "Shut up, rookie. It's because we have always done it that way," "I don't know, whatever, it's none of your business," "Who cares," and so forth. By the way, the department in the example above no longer washes truck tires every night.

Organizational culture is important because it represents who we are. Moving forward, we must take care to cultivate the right organizational culture. We must encourage those around us to ask why, to ensure that we are providing the right services in the right ways.

## Complacency

Complacency is a relaxed attitude toward operations that is commonly based on successive experiences where nothing ever went wrong. It can be taught, and it can be learned—sometimes quickly. With years of reinforcement, complacency becomes part of the organizational culture and becomes a deeper version of itself called "normalization of deviance"—that is, where the new lower standard of performance becomes the "normal" standard that is now taught to new firefighters. CRM places a strong focus on fighting complacency because it is often one of the root causes contributing to accidents. The following example illustrates this.

### Case in Point

A driver (or engineer, apparatus operator, or chauffeur) shows up every day at the station and checks the truck. The driver checks the lights, fluids, systems, and equipment thoroughly. After a few months or a few years, the driver figures it is a waste of time to check inside every medical bag or toolbox because the contents are always the same. The driver also stops checking every piece of equipment. This is complacency.

Over the next few years, this driver teaches several new drivers how to check the truck. He tells them there is no need to check in such detail inside every medical bag and toolbox. This is normalization of deviance. This department now has a new, lower standard of operation. As it moves through several drivers, it becomes part of the department's culture. Complacency is a killer because it sneaks in over time, making it difficult to detect and combat.

Combating complacency requires a continuous focus on pride and performance. It requires continuous work to ensure that all operations are maintained at the highest possible levels. This is another tenet of CRM: continuous focus, work, and practice in a process that never ends. Real solutions have no end; they are ongoing.

Methods to combat complacency include:

- Teaching crew members to connect with and focus on something larger than themselves:
  - Service to the community
  - Success of the crew, shift, department, or profession
- Encouraging pride in performance
- Reviewing near-miss reports to learn from how things have gone wrong for others
- Refusing to ever let your performance of duties slip regardless of fatigue or lack of time
- Teaching others to fight complacency

*When it comes to making mistakes, there are two types of people: those who have and those who will. The moral of the story is: be careful about pointing fingers.*

The above phrase has been overheard in reference to people flying aircraft with retractable gear who have made a gear-up landing—that is, a landing where the pilot simply forgot to put the landing gear down. It is intended to illustrate to people who think it will never happen to them that unless they truly learn from others' mistakes and remain focused on pride in performance, they too will make the same mistakes. The moment you think it will never happen to you is when you are really in trouble. It shows the attitude many pilots carry with them to fight complacency. They constantly tell themselves that if they are not focused and careful, they will be the next to land gear up.

This same line of reasoning applies directly to the fire service. Examples include incidents where the pump operator charges a line while it is in the hose bed or a member of the crew forgets a piece of PPE or to refill the apparatus water tank. Now, apply the quotation where it is applicable to your own operations.

## Trust versus loyalty

For what we have talked about—namely, CRM—to work, you need to build trust. We learned the importance of trust earlier, in the discussion of human factors (see chap. 1). Applying CRM means trust must exist within crews, but when it can be facilitated among an entire station, shift, or department, that is even better. Trust is the key ingredient in individual, crew, departmental, and fire service growth because it facilitates sharing and learning. Where there is no trust, there is no sharing. If there is no sharing, then we cannot learn from the experiences and mistakes of others. Moreover, we cannot expose true weaknesses in our operations because they will stay hidden to avoid embarrassment, punishment, or ridicule. Trust encourages open communication.

Unfortunately, many of us get trust and loyalty confused. Loyalty appears to be beneficial at first glance, but it is problematic if it is built through hazing. Loyalty developed from hazing requires a person to do something or put up with something (often embarrassing and humiliating) as a requirement for membership (e.g., to a crew, club, department, or group). During this type of loyalty-building process, any trust is completely destroyed; free thinking and sharing are also eradicated. After someone just embarrassed you in front of others, do you feel like trusting them or sharing? The downfall of this is that an individual who is loyal follows the group regardless whether they are right or wrong. This is not what we are looking for in the modern fire service. In the modern fire service, we are looking for individuals who use their brains to lead, assist, evaluate, participate, and mitigate professionally and ethically—not just to be yes-men. Herein lies the problem with loyalty. Although often well intended, it actually erodes trust and prevents critical thinking when used improperly, which are critical to safety and success.

> *Without trust, we cannot move forward through learning as individuals, crews, departments, and a fire service.*
>
> —*Craig Nelson*

# ERRORS

Another problem area that CRM addresses is the realm of errors. For the sake of this discussion, errors can be broken down into two categories: purposeful and accidental (table 8–1). Have you ever stopped to think about what happens after someone commits an error? When should you punish them? Why do we punish errors? What is the goal of punishment? In the context of a learning environment, punishment should be used only when an error is purposeful, not when it is accidental. Very few errors in the fire service are intentional, so punishment is overused, given that it works well only in purposeful cases.

**Table 8–1.** Distinction between an error versus a violation, and how an error should be handled (Source: Firefighter Near Miss Resources 2013)

| Error | Violation |
|---|---|
| Lack of skill; lack of education or training | Willful disregard for rules and regulations |
| Poor decision making; misperception | Willful disregard of previous retraining |
| **How Errors Should Be Handled** | **How Errors Are Often Handled** |
| You are human | You are highly trained |
| Humans make mistakes | If you did as trained,<br>you would not make mistakes |
| Find the root cause | You weren't careful enough |
| Improve the system | You should be punished |

In accidental cases, punishment often escalates the situation to make it worse. Nevertheless, it is used because it is quick and easy—even though it usually is not a true solution. Consider a driver who accidentally backs into a post; punishment may fit the department's policies, but it also may reduce confidence and it does not prevent the driver, or any other driver, from doing it again. Punishment also does not prevent someone else from making the same error again. The behavior is more effectively changed in the driver and everyone else when an investigation is done to reveal the root cause(s), when operations are adjusted accordingly, and when training is developed to teach everyone about the operational adjustment. When we share what was learned from the event and train everyone, we reduce the odds of it happening again. This is the goal—fewer accidents and fewer injuries.

## Ending the blame game

As discussed earlier, developing a learning culture is about creating a crew and departmental movement focused on improvement. CRM promotes a learning culture to improve safety, reduce accidents, and improve operations. Get your people reading, get them addicted to learning and self-improvement, get them training.

A learning attitude and trust develop the right environment to begin incident review and near-miss reporting. Remember from earlier that incident review can be as simple as sitting down together after an incident for a tailboard talk. The purpose is to discover what was done well (behaviors that should be repeated) and what could have been done better (behaviors that should be changed). Near-miss reporting is the process of sharing and collecting in one place the lessons learned from the incident review process so that everyone can benefit. It entails sharing something that could have gone or did go bad with others so that they can learn from the experience without paying the price of experiencing it themselves. The point is to learn collectively as a fire service profession from the past in an effort to predict the future and then develop training to address potential incidents before they happen again.

Key ingredients to a good review are:

- It should happen as soon as possible after an incident for accuracy
- It must happen in a trusting environment for people to admit mistakes and to obtain useful information and data
- Its intent is learning, not blame or punishment
- No Monday-morning quarterbacking (it is easy to say what you would have done differently after the incident is over)
- The more incidents you review (your own, your department's, and those from other departments), the more you will learn

It is important to complete a root-cause analysis to discover what causes were truly involved with a near-miss, incident, or accident. One method for doing this is to build a chain of errors as discussed earlier (see chap. 4). Remember that if any of the links (causal pieces) are removed, then the accident is prevented.

Key ingredients to assist with root-cause analysis are:

- It should move beyond symptoms to fundamental behaviors
- It looks for all the factors contributing to an incident that may have surfaced days, weeks, or months prior
- It places the factors into a chain of errors
- An analysis of the causal factors occurs
- A recognition that the removal of any link in the chain of errors prevents a specific incident
- To develop training or operational changes to address the root cause to prevent similar incidents in the future

## Tracking and trends

Incident review and a near-miss reporting system makes possible the collection of useful data. The more data (incident reports) collected, the more useful and accurate the information will be about what is currently causing and will continue to cause problems. After collection of enough data, it becomes possible to identify trends (problems). The important issues to look at are those that are most common and those with the most severe consequences. An exercise illustrates this best.

### Exercise 8–5

As a small group in your department, as a crew, or on your own, list the incidents where you or others in your group have witnessed an operational issue within the categories in the following list. Place a tally by each category for the number of times it has been witnessed as an issue. Here are several common issues to start the list:

- Apparatus operations
- PPE
- Rehab
- Communications
- Ladders

- Hoselines
- SCBA
- Ventilation
- Tactics
- Command
- Situational awareness
- Contributions of other entities (e.g., police, EMS, public works, utilities)

By tallying each category, you will see quickly, on a basic level, where you, your crew, and your department need to focus future training efforts. For example, if apparatus operations receive the most tallies, it is an area that needs more focus in training. If you formalize this across your organization, you will begin to gather accurate data about your current and future operations, but this is contingent on trust, incident review, and incident reporting all taking place.

---

## LEADERS, FOLLOWERS, AND TEAMWORK

A critical area of CRM is individual leadership. This is a term that we hear often and a term that has been discussed previously. However, CRM is heavily reliant on effective leadership. Many people present classes about it. Many books have been written about it. But what is it? What have you been taught that it is? When you actually stop and think about it, the concept is relatively complicated for a term that is thrown around so often. Leadership is . . . well, maybe it is easier to explain such an intangible term with a list of traits; thus, leadership is:

- Openness to change
- Agreeableness
- Conscientiousness
- Extraversion

- Vision
- Attitude
- Morality
- Influence
- Independence
- Hard work
- Fairness
- Consistency
- Teaching others

Two key parts of leadership directly related to CRM include group development and understanding how the leader affects the group's communication and decision making. These two topics have also been discussed heavily; nevertheless, the following short recap is provided because they are so important to CRM.

## Group development

Groups (crews) take time to develop. They take time to develop their set of norms—behaviors—that reflect the disposition, attitude, and organization of the group. As a group develops, those in it begin to find their roles. Remember, it takes six months to a year for a group to develop to the point where the roles are defined. After this initial period, those in it have learned their roles, their duties, how the group will function, and the group's values. This is why crews that have worked together for a long time often perform better than a crew that has just been assembled. Individuals within the group that has worked together know not only their own roles but also those of the other crew members, much like a professional athletic team that has played together for years. They know who is doing what, when they are doing it, and how they are doing it—allowing all members to smoothly work together.

## Knowing yourself as a leader

The leader is the one who shapes the group's development, so it is important as a leader that you know yourself. It is important that you consider how others perceive you, because these perceptions shape how they behave to meet the leader's expectations as they understand them. In our own minds, we always think of ourselves as great people: as good listeners, approachable, fun, hard-working, and so forth. However, others often do not always see us in such rosy terms, and the truth may lie somewhere in the middle. Regardless, we need to know what our own strengths and weaknesses are—honestly. Free assessments are available online that give general guidance (search for "leadership style" or "personal inventory").

## PUTTING CREW RESOURCE MANAGEMENT TO WORK

Implementing CRM is an important step, but realize that it is a change—and as discussed previously, changes invoke certain reactions in all humans. Thus, CRM must be introduced (just as any other organizational change would be) in a thoughtful manner.

CRM must first be explained. Explain, or have someone familiar with it explain, what it is and why it is important. After the seed has been planted, a formal classroom presentation on CRM will give members the tools necessary for a basic understanding of the concepts and terminology. They will learn how to begin applying it.

The following are keys to successfully implement and sell CRM:

- Recognize that fire service resistance is rooted in culture and identification with the past.
- Describe how CRM strengthens authority in leadership instead of weakening it.
- Show data that support CRM and explain that your community, media outlets, and politicians, among others, expect data-driven decisions.
- Remember that data-driven decisions are defensible and show how data-driven decisions can help with gaining support for the future.

- Gain acceptance by whole organization.
- Provide training in and employ CRM across the entire organization.
- Implement CRM in small steps because it will work at company level and can grow from there.
- Practice CRM during realistic, scenario-based training to bring out desired behaviors and skills.
- Acknowledge that CRM is a constant process; it is never finished.
- Develop a personal learning attitude.
- Support an organizational learning culture.

Build the CRM system by adding incident review, incident reporting, and incident tracking after introducing the basic concepts. Then, identify trends, select areas for increased training, and adjust the training as data are collected. Training must be designed to bring about the behaviors that need to be addressed. For example, if not wearing PPE is an issue, then the training exercise must be built to force the crew into situations where they will practice wearing their PPE properly. For training to have the greatest impact, it should be scenario based and designed to simulate realistic conditions as closely as possible. In the PPE example, talk with the crews about the problem and then have them put on their PPE properly in the station. The next step is to take them out to the training grounds and present or simulate fire conditions. Here, they need to perform all of the other functions under more realistic conditions at the same time that they are being monitored for their correct PPE use.

One method to monitor the effectiveness of CRM once it becomes part of normal operations is a line operations safety audit. This involves a training person riding along with crews to monitor how many threats they experience during incident response. Threats are outside forces that could cause an incident or accident. An example of a threat would be a fire truck traveling through a green light when another vehicle crossing the intersection fails to stop on their red so the car almost hits the truck. The training person also analyzes errors that are made and tracks whether the errors are caught and corrected by the crew working together.

In a line operations safety audit in the airline industry, it was found that:

- 98% of all flights face one or more threats.
- There are on average four threats per flight.
- There are errors on 82% of all flights.
- There are on average 2.8 errors per flight.
- Nearly all flights end safely and successfully owing in large part to CRM because crews working together catch the errors and threats and address them before they develop into serious problems.

Airline flights are well-planned events that still encounter several threats and errors. Given that, consider the potential for threats and errors in an unplanned emergency response. CRM is credited with preventing the threats and errors from becoming accidents by improving the crew's ability to deal with them.

## REFERENCES

Besco, R. D. n.d. To Intervene or Not to Intervene? The Co-Pilot's Catch 22. http://picma.org.uk/sites/default/files/Documents/Background/Besco%20 Co-pilots%20dilemma.PDF.

Carley, D. A. 2013. *Communication Impediments and Support in the Fargo Fire Department's Organizational Learning Process*. Emmitsburg, MD: National Fire Academy.

Firefighter Near Miss Resources. 2013. Retrieved from firefighternearmiss.com.

Gladwell, Malcolm. 2008. *Outliers: The Story of Success*. New York: Little, Brown and Co.

International Association of Fire Chiefs. 2003. *Crew Resource Management: A Positive Change for the Fire Service*. 3rd ed. http://www.iafc.org/ files/1SAFEhealthSHS/pubs_CRMmanual.pdf.

Klein, G. A. *Sources of Power: How People Make Decisions*. 1998. Cambridge, MA: MIT Press.

LeSage, P., J. T. Dyar, and B. E. Evans. 2011. *Crew Resource Management: Principles and Practice*. Sudbury, MA: Jones and Bartlett.

# CONCLUSION

The concepts discussed in this book are big ideas. Some of them come from within the fire service, whereas others come from other industries that share similar traits. Each concept or idea represents a piece to a puzzle—and the HRO puzzle cannot be complete without all of them coming together. Regardless of where the idea comes from, it takes time to develop it and see it practiced. If a fire department decided today to implement the concepts presented here, it would take years—probably a generation—to see a strong evolution toward change.

This is for two reasons. First, an HRO never finishes learning, so there is no end point. But more important, it takes time to develop the behaviors and practices in a department and to develop the people within the department. That is one reason why the chapter on change is so important. If handled incorrectly, the change may never happen. However, the fire service has always adopted change that benefits the department and the community in the long run.

Leadership and the tools that help us to identify the root cause(s) of problems will benefit the fire service and those who want to see the fire service excel. The behaviors and principles exhibited by its leaders will benefit not only the fire service but also—and at least as important if not more—the communities each department serves.

In many cases, the fire service already does many of the things described in this book. However, the intent of outlining all of this is to provide us with a new direction to build or use systems to their maximum potential. It helps us to repeat our successes and improve on incidents that are not as successful.

The fire service is filled with many passionate and talented people who have done great things. Because of them, the fire service will continue to do great things, but only if we continue to be open minded. We must continue to place the well being of the communities we protect first.

Our hope is that this book can be used as a tool to help anyone at any level of the fire service to bring about changes—to spark discussions, to advocate looking outside of the fire service for ideas, to encourage people to strive. Ultimately, this is how we expect to make ourselves, our crews, our departments, the fire service, and our communities better and safer.

# A

# SOLUTIONS TO EXERCISES AND DISCUSSION QUESTIONS

## Discussion Question 1–1

Some ideas to help start the list for this discussion question are:

- Physical
- Enjoy the outdoors
- Enjoys riskier activities (e.g., motorcycle riding, physical sports)
- Teasing
- Dark humor
- Desire to help others
- Team player

## Discussion Questions 1–2

Everyone has been overwhelmed at some point during an incident. It may have been when you were new in the fire service or new in a position. Did you handle it by relying on training? Following the orders you were given? Following a senior officer? Asking for input? Reducing your workload?

## Discussion Question 1–3

A common example of this in the fire service is the fog-versus-solid-stream argument. Debates should be based on science, but often they are based on a small sample of personal experiences or emotion.

## Discussion Questions 1–4

People react to what they are hearing before they have time to process it completely, so this is more common than it may seem. It helps to calm down, wait, and think about what is being said instead of forming a response before the other person is done talking (remember how Mom always said to count to ten before replying?).

## Discussion Questions 1–5

Reasons someone in the fire service may go with the group:

- Lack of seniority
- Willingness to fit in
- Being unsure if one's idea was more correct than another's
- Fear of speaking to or contradicting an officer

It is typically hard to speak up in the fire service for these reasons, among others. It is important to overcome this so that all of our crew members actively participate in problem solving.

## Discussion Question 1–6

Here are some additional suggestions:

- Personal issues (e.g., kids in school, money, divorce)
- Heavier traffic than normal when responding to a call
- Fatigue (e.g., excessive partying, new baby at home, working long hours)

## Discussion Questions 1–7

Most likely, everyone exhibits one of these traits during their career. The key to overcoming them is understanding how they affect the crew's performance. Once this is understood, the traits can be better explained to the person exhibiting them. It is also important to understand why these traits are being exhibited so that the root cause(s) can be addressed.

## Discussion Questions 2–1

Hopefully people in your department do ask why; still, because of our hierarchical system, it is often not asked as often as it should be. It is a great sign if people in your department are asking why because it is a strong indication that a learning culture exists. Stefan Svensson says it took the Norwegian fire service 15 years to build a learning culture that is not afraid to ask why. This culture started in the training environment. Because it takes time, the sooner asking why is built into the environment, the better.

## Discussion Question 2–2

The how tool is important because it puts the ideas that come from asking why into operation. In other words, it is where the rubber meets the road.

## Discussion Questions 2–3

We have heard solid arguments saying that human factors can influence or be influenced by all of the categories. For example, humans write SOGs; thus, if an SOG played a role in a near-miss, then there was a human factor involved. Equipment may fail because it simply failed, or it could have failed for a lack of required maintenance, which would also be a human factor.

## Discussion Question 2-4

Many departments focus their efforts on technical skills. While it is important to focus on technical skills, it is just as important to focus on the human side of the skills at the same time.

## Discussion Questions 2-5

Possible answers include:

- Teamwork—a new crew member was not aware of the typical operations
- Procedure—one member was used to doing something one way while another was used to doing it differently
- Staffing—moving crew members around to meet staffing demands meant having a different person filling in than normal
- Communication—not communicating intended actions between two people involved led to one not knowing what the other was doing
- Situational awareness—lack of communication contributed to poor situational awareness

## Discussion Question 2-6

You learned not to do "it" again by being punished if it was something done intentionally. However, if it was unintentional the first time, you now lose confidence not knowing if it can prevented in the future. Plus, you are frustrated because the department singled you out. This often leads to one of the seven hazardous attitudes, such as resignation. The other side of it is that departments often miss a learning opportunity when something is done unintentionally, so that others are taught not to do the same thing in the future.

## Exercise and Discussion Questions 3–1

Some possible answers (in addition to those in fig. 3–1) include:

- Person is task oriented
- Person follows the rules (e.g., SOGs/SOPs)
- Person believes seniority deserves its advantages over all else

A person who is a manager can be successful by following the rules and doing what is expected. Such people help the department by completing objectives, tasks, and training, among other things.

## Exercise and Discussion Questions 3–2

1. Possible answers (in addition to those given in fig. 3–1) include:
   - Outcome and/or personnel oriented
   - Best practices
   - Merit deserves advantages
2. Common strengths we have seen in firefighters include:
   - Problem solving
   - Being go-getters
   - Team building
3. Common weaknesses we have seen in firefighters include:
   - Not always being open to new ideas
   - Not always being good listeners
   - Tending to focus on what we are good at and avoid things we are not as good at
4. Your description of a leader hopefully includes some of the terminology and ideas in the chapter—such as being open to new ideas, good at communicating, and considering others' positions before responding.

5. Hopefully your description does not focus on rank, title, or seniority.

## Exercise and Discussion Questions 3–3

1. Common leadership qualities include honesty, integrity, adaptability, openness to new ideas, ability to motivate, and compassion, among others.
2. Your department may or may not train on and test for some of these traits. The typical fire service testing process does not assess leadership abilities but instead tests rote memorization. Leadership can be taught, and a person's leadership ability can be measured as discussed in the chapter.

A fire service testing process is typically weighted heavily toward rote memorization and technical skills. Written tests and assessment centers tend to be designed to measure these abilities and have little to do with leadership. Although interviews are an important part of the process, they are considered unreliable in predicting work performance, so it is unlikely that most interview processes measure leadership. None of these processes are incorrect, but it is important to be sure the processes are actually measuring for the behaviors and traits the department wants in the person being assessed.

## Exercise and Discussion Questions 3–4

As described in the book, it typically takes six to twelve months for a group to develop. This exercise helps you identify how far your group or crew has made it in the formation process. Is it progressing appropriately? Did it stick in one stage or another? Now you can move it forward and shape it if it did, or you can maintain the current status if the crew is progressing appropriately. Other groups you may have been a part of in the past could be athletic teams, church groups, and clubs, among others.

## Exercises and Discussion Questions 4–1

Increased use of the near-miss reporting system may happen with:

- Increased recognition of the system in recruit training
- Dispelling myths about identification and punishment (because it allows anonymous reporting)
- Rewarding use of the system
- Explaining the value of the system

Some other factors include:

- Lack of time
- More work
- Unfamiliarity
- Not understanding the value of such a system

## Discussion Questions 4–2

In our experience, the fire service tends to focus on the 20% that is a technical fix to a problem, rather than on the 80% that is a human factor–based solution. Continuing to focus our efforts on the 20% provides us with diminishing returns because only so many fixes can happen there. The 80% provides the most opportunity for improvement.

## Exercise 5–1

This exercise is about reflecting on your department's progress and ability to adapt to a changing environment or changing expectations. Review the list above the exercise. Is your department part of these changes? Is your department part of changes not on the list? What could the department explore in the future?

## Discussion Questions 6–1

The lessons learned in the Real-Life Example provide many connections to the bullet points that precede the example and summarize the principle. Most, if not all, of these points are represented in this example. Examples of other behaviors fitting this principle that may be added include lack of communication between the firefighter and the shift commander and complacency.

## Discussion Questions 6–2

1. The culture contributed to this event because the crew did not question or clarify orders that were not specific or clear enough.
2. Links that may be in your chain of errors starting at the incident and working backward include: confusion at the outset because of a lack of address; second crew is delayed; started fighting fire above before fire on the ground floor was controlled; unclear assignment; lack of accountability; no established RIT; lack of situational awareness; lack of assertiveness; lack of communication training; and fire service hierarchy. Your chain of errors may be slightly different. Did your chain of errors lead past the operational items and begin exploring the cultural items given in the list above?
3. Questions could have been asked at three points: first, when command ordered the crew to pull a line around the back, because "the back" could mean different places to different people, depending on where they were standing; second, when the crew could have asked where the second line was and/or when it would be in operation; and third, at the time of establishment of a RIT, if there was one. At any of these three points, clarification through questioning could have changed the outcome. Unfortunately, it took a Mayday situation to cause the officer to go to the incident commander and confront the situation, by suggesting a different strategy only after the Mayday had already been declared.

## Discussion Questions 6–3

1. Completing repetitive simple tasks can lead to us completing them mindlessly; with no focus on the task itself, each task is reduced to a check on a checklist. Mindless routine leads to complacency. Eventually, complacency leads to a normalization of deviance—the acceptance of doing things less than ideally. In this example, it can lead to a dead battery or the wrong channel on the portable when it becomes normal to simply check that the portable is there without checking its operation. There are many other routine tasks we do each day that can contribute to similarly detrimental outcomes when there is normalization of deviance.

2. The RIT had excellent situational awareness because what they thought was happening matched what was actually happening. On the other hand, the attack crew was in the opposite situation. Because the crew with situational awareness was unable to communicate with the crew losing situational awareness, this could not be improved. This shows the importance of constant communication.

3. Understanding how the thermal imager works helps the user to understand its shortcomings and strengths. Unfortunately, many departments obtain their training from a short session with the vendor in a fire station, so the operational aspects of their use are often not well understood.

4. Instead of arguing over which is more important, this shows that both experience and education are important. Experience is the application of knowledge, whereas knowledge helps us understand the why in the experience. The two together gives us a fuller picture of what is actually happening, allowing us to operate at a higher level.

## Discussion Questions 6–4

1. The interaction between the officer and the incident commander that illustrates the action of listening to those on the streets is when the officer advised the incident commander to maintain units on scene because there was a fire that they had not found yet. The incident commander could react effectively because there were two prepositioned lines.
2. The incident commander assumed that the light haze was not going to become much of a problem. That, combined with the information for the other run, likely contributed to the assumption that the second run was more important. The root cause may have come from some complacency with the situation. A lot of departments may see this when first-arriving officers or incident commanders release apparatus before they have fully investigated the situation.

## Discussion Questions 6–5

Consider the people on your crew and their skills. For example, framers will have a solid understanding of building construction, and electricians will be helpful at an incident involving electricity. These skills should be—and usually are—used to the crew's advantage.

## Discussion Question 6–6

An aide can come from mutual aid; be a driver, operator, or engineer; be an incoming officer from another company; be dispatched as part of an automatic aid arrangement; be another chief officer.

## Discussion Questions 6–7

1. There may have been too much information for one radio channel or for one person to absorb. However, information helps everyone develop situational awareness. So what are the solutions? We would suggest a second channel and an aide

to help listen to the traffic. We would also suggest training in concise communications so that the same information could be transmitted in shorter bursts. Given that, we do think a general solution is to reduce the volume of communication without losing the content. It is a matter of training and adapting the system in place to prevent this from happening.

2. We strongly believe that should not be the only time. Any time any fire personnel on scene have relevant information that will increase situational awareness, affect the outcome of the scene, alert to changing conditions, or serve a resource need, it should be transmitted in concise messages. This does not mean that everyone should put everything on the radio. It needs to meet the four criteria just listed.

3. We do not know of a way for the incident commander to gather information effectively without information traveling up the chain of command. The incident commander is limited to an exterior view, so the information gathered is incomplete.

4. Using the previous command aide exercise, consider the best way to add a command aide to your operations.

## Discussion Questions 6–8

1. It is showing initiative if the actions taken are communicated up the chain of command. A lack of communication is considered freelancing. Heroism should not even be part of the equation. The outcome should not be the determining factor in how an action is characterized. Instead, the determining factor should be whether the communication was appropriate. A person taking action and making a successful rescue without communicating it is still freelancing. Conversely, a person who takes action that is unsuccessful but communicated the actions taken is showing initiative.

2. In this situation, we would have done the same thing and ensured that it was communicated up the chain of command. It should not change how the action taken is labeled.

3. It is not freelancing if the actions taken are communicated to others.

## Discussion Questions 6–9

1. The responders retreated three times before resolving the situation.
2. Only the original patient was hurt, with two crew members having a near-miss because of the appropriate way the incident was handled.
3. The original victim plus the entire crew on the first-responding apparatus could have been hurt.
4. The firefighters came out of the basement when they realized it was an unsafe environment; the captain pulled everyone back because he was unsure of the situation; a hazardous materials response was requested; and coming back out of the environment when the lower explosive limit was reached are examples of initiative in the case study.
5. The answer to this is specific to your department.
6. The answer to this is specific to the incidents you chose in question 5.

## Discussion Questions 6–10

1. Administrative examples include building stations as call volume grows, employing an increasing number of part- or full-time firefighters as volunteerism drops, and adding services such as EMS or hazardous materials response.
2. This depends on your department. However, examples include teaching others to consider alternative solutions, openly evaluating new ideas, and encouraging everyone to provide ideas.
3. It is likely that the vast majority of fire departments lack diversity because we have low turnover and tend to promote from within. Furthermore, we do not attract a wide variety of people to our profession. Even though this stability has benefits, it also makes it harder for us to adapt and prepare for the future.

## Discussion Questions 6–11

1. Although this question is specific to the reader and crew working on these questions, the goal is to help us realize how often we adapt to an evolving situation.
2. Although this question is again specific to the reader and crew working on these questions, the goal is to help us realize how this concept applies to administrative decision making as much as it applies to operational problems.
3. The can-do attitude is absolutely necessary for the fire service to be successful. However, it is just as important for our energy and know-how to be focused on goal attainment. Not focusing energy can lead to "moth to the candle" syndrome, machismo, and poor decision making.
4. Diverse backgrounds bring a different perspective to a problem, both operationally and administratively. A lack of diversity leads to status quo and groupthink. However, too much diversity leads to indecision and lack of direction.
5. Initially, the first-arriving firefighters adapted to their situation by trying different tools to open the hood. However, the first-arriving firefighters focused on opening the hood at the expense of using a temporary measure to knock down the fire while working on the lock. This case study illustrates both adaptation and the importance of having a diverse background represented within a department or crew.

## Discussion Questions 6–12

1. The different responding agencies could have trained together ahead of time either through tabletop scenarios or on developing shared terminology. This would have led to a smoother response because EMS would have recognized, through familiarity with the other agency's operations, that the officers' drawn weapons were a sign of a potentially dangerous situation.

2. Examples include law enforcement; EMS; dispatch; city, county, or state agencies; bus companies; utilities; school districts; and construction companies.

## Discussion Questions 6–13

1. No, they should not, because we cannot possibly create an SOG for every situation that we may ever encounter in the fire service. When we attempt to do so, we are going beyond basic guidelines to attempting to create a rule for every situation.
2. No, because they are not meant to address every specific situation. They are meant to be general guidelines only, to help crews make their own specific decisions addressing each situation. It is not possible to write an SOG for every situation.
3. Discipline does not usually help a crew member who has made a mistake. Instead, it often drives crew members away and undermines their confidence, leading to more problems. It also does not prevent someone else from making the exact same mistake, which may be even more serious when it happens again in the future.
4. A value-driven solution means teaching people to make good decisions and react appropriately. When people in an organization are taught the values, vision, and mission of the organization, they have the tools they need to make good decisions in unusual circumstances. A list of rules for what to do and not to do is not necessary and often only complicates the scene, because people expend energy and time trying to remember what they can and cannot do.

## Discussion Questions 8–1

1. Much of our information tends to come from within our own organizations. It helps an organization be successful to look at sources from outside the department, such as trade journals, textbooks, conferences, and other industries for methods, processes, or technologies that can be useful in our industry.

2. It is important when gathering information to choose reputable sources. However, it is impossible to know for sure. One key method is to gather information from more than one source or from a source that incorporated and documented broad-based research methods.

3. Other industries with similar traits include EMS, law enforcement, airlines, military, emergency medicine, oil drilling, or any service-oriented business that has developed repeatable, successful processes.

## Exercise 8–1

Some traits that the listed occupations share with the fire service include:

- Law enforcement: Hierarchical structure, emergency response, boredom interrupted by emergencies
- Military: Hierarchical structure, high-risk environment
- Airlines: Hierarchical structure, work as crews, boredom interrupted by emergencies
- Education/teaching: Service oriented, education oriented
- Emergency rooms: Hierarchical structure, time compressed, life safety factor, health service
- Air medicine: Hierarchical structure, work as crews, boredom interrupted by emergencies, time compressed, life safety factor
- Service-oriented business: Service oriented, customer satisfaction important, services meet the needs and expectations of the customer

## Discussion Questions 8–2

1. Typically, these statements are little more than well-intended gestures. Although they may contain deeper meanings, they mean little at face value as they are often used.
2. Realizing the deeper meaning of these sayings is accomplished by:
   - Identifying true safety issues, rather than only the symptoms
   - Identifying root causes of our accidents
   - Understanding what having situational awareness actually means
   - Understanding what affects good decision making
   - Developing an environment that encourages open communication

## Exercise 8–2

Examples of flags include:

- Incident involves children
- No safety officer
- New crew members
- Something does not make sense
- No RIT
- Stuttering or other speech abnormalities
- Withdrawing from the incident
- Unusually high amount of stress from personal life (e.g., divorce, terminal illness in immediate family, financial hardship)

## Exercise 8–3

Examples of ways to gather information to gain situational awareness include:

- Mobile data computer/mobile data terminal (MDC/MDT)
- Asking crew members for insight
- Completing a walk around
- Talking to bystanders
- Preplanning
- Increasing communication
- Using an aide
- Reducing stress
- Calling for additional resources

## Exercise 8–4

*Probe:* "Captain Stevens, I'm concerned that the roof is going to collapse because of multiple reports from the truck company. Would you agree that we need to exit?" Wait for a response from the captain.

*Alert:* "Captain Stevens, I'm concerned that the roof is going to collapse because of multiple reports from the truck company. We need to exit." Wait for a response from the captain.

*Challenge:* "Stop, Captain Stevens! The roof is going to collapse so we need to exit now!" Either wait for a response from the captain or observe from his actions whether he is in compliance with your suggestion.

*Emergency intervention:* As you grab Captain Stevens to pull him off the third floor, you say, "We are leaving right now."

## Exercise 8–5

Look at the operations and tasks with the most tally marks and the ones with most severe consequences. These are the operations and tasks that need the most the most focus during training in the future.

# INDEX

## A

## B



## C

## G



## H

## I

# J



# K



# L

# M



# N

# U



# V



# W



# Z